Mathematica®
in Theoretical Physics

Selected Examples
from Classical Mechanics
to Fractals

Gerd Baumann

Professor Gerd Baumann
University of Ulm
Albert-Einstein-Allee 11
89069 Ulm
Germany

This is a translated, expanded, and updated version of the original German version of the work "*Mathematica*® in der Theoretischen Physik," published by Springer-Verlag Heidelberg, 1993 ©.

Publisher: Allan M. Wylde
Publishing Associate: Keisha Sherbecoe
Product Manager: Walter Borden
Production Management: Black Hole Publishing Service

© 1996 Springer-Verlag New York, Inc.
Published by TELOS, The Electronic Library of Science, Santa Clara, California
TELOS is an imprint of Springer-Verlag New York, Inc.

Cataloging-in-Publication data is available from the Library of Congress

Printed in the United States of America

9 8 7 6 5 4 3 2 1

ISBN 0-387-94424-9   Springer-Verlag   New York   Berlin   Heidelberg

TELOS, The Electronic Library of Science, is an imprint of Springer-Verlag New York with publishing facilities in Santa Clara, California. Its publishing program encompasses the natural and physical sciences, computer science, economics, mathematics, and engineering. All TELOS publications have a computational orientation to them, as TELOS' primary publishing strategy is to wed the traditional print medium with the emerging new electronic media in order to provide the reader with a truly interactive multimedia information environment. To achieve this, every TELOS publication delivered on paper has an associated electronic component. This can take the form of book/diskette combinations, book/CD-ROM packages, books delivered via networks, electronic journals, newsletters, plus a multitude of other exciting possibilities. Since TELOS is not committed to any one technology, any delivery medium can be considered.

The range of TELOS publications extends from research level reference works through textbook materials for the higher education audience, practical handbooks for working professionals, as well as more broadly accessible science, computer science, and high technology trade publications. Many TELOS publications are interdiscilinary in nature, and most are targeted for the individual buyer, which dictates that TELOS publications be priced accordingly.

Of the numerous definitions of the Greek word "telos," the one most representative of our publishing philosophy is "to turn," or "turning point." We perceive the establishment of the TELOS publishing program to be a significant step towards attaining a new plateau of high quality information packaging and dissemination in the interactive learning environment of the future. TELOS welcomes you to join us in the exploration and development of this frontier as a reader and user, an author, editor, consultant, strategic partner, or in whatever other capacity might be appropriate.

TELOS, The Electronic Library of Science
Springer-Verlag Publishers
3600 Pruneridge Avenue, Suite 200
Santa Clara, CA 95051

TELOS Diskettes

Unless otherwise designated, computer diskettes packaged with TELOS publications are 3.5″ high-density DOS-formatted diskettes. They may be read by any IBM-compatible computer running DOS or Windows. They may also be read by computers running NEXTSTEP, by most UNIX machines, and by Macintosh computers using a file exchange utility.

In those cases where the diskettes require the availability of specific software programs in order to run them, or to take full advantage of their capabilities, then the specific requirements regarding these software packages will be indicated.

TELOS CD-ROM Discs

It is also clearly indicated to buyers of TELOS publications containing CD-ROM discs, or in cases where the publication is a standalone CD-ROM product, the exact platform, or platforms, on which the disc is designed to run. For example, Macintosh only; MPC only; Macintosh and Windows (cross-platform), etc.

TELOSpub.com (Online)

New product information, product updates, TELOS news and FTPing instructions can be accessed by sending a one-line message:

> **send info**
> to:
> **info@TELOSpub.com.**

The TELOS anonymous FTP site contains catalog product descriptions, testimonials and reviews regarding TELOS publications, data-files contained on the diskettes accompanying the various TELOS titles, order forms and price lists.

To Carin,

for her support and encouragement.

Preface

As physicists, mathematicians or engineers, we are all involved with mathematical calculations in our everyday work. Most of the laborious, complicated, and time consuming calculations have to be done over and over again if we want to check the validity of our assumptions and derive new phenomena from changing models. Even in this age of computers, we often use paper and pencil to do our calculations. However, computer programs like *Mathematica* have revolutionized our working methods. *Mathematica* not only supports popular numerical calculations but also enables us to do exact analytical calculations by computer. Once we know the analytical representations of physical phenomena, we are able to use *Mathematica* to create graphical representations of these relations. Days of calculations by hand have shrunk to minutes by using *Mathematica*. Results can be verified within a few seconds, a task that took hours if not days in the past.

The present text uses *Mathematica* as a tool to discuss and to solve examples from physics. The intention of this book is to demonstrate the usefulness of *Mathematica* in everyday applications. We will not give a complete description of its syntax but demonstrate by examples the use of its language. In particular, we show how this modern tool is used to solve classical problems.

This book contains a large number of examples that are solvable using *Mathematica*. The defined functions and packages are available on a disk accompanying this book. The names of the files on the disk carry the names of their respective problems. Chapter 1 comments on the basic properties of *Mathematica* using examples from classical mechanics, especially problems involving the oscillator. Chapter 2 demonstrates the use of *Mathematica* in a step-by-step procedure. We show how this software is used to simplify our work and to support and derive solutions for specific problems. In Chapter 3, we discuss problems of electrostatics and the motion of ions in an electromagnetic field. We further introduce *Mathematica* functions that are closely related to the theoretical considerations of the selected problems. Chapter 4 discusses problems of quantum mechanics. We examine the dynamics of a free particle by the example of the time–dependent Schrödinger equation, and study one dimensional eigenvalue problems using the analytic and numeric capabilities of *Mathematica*. In Chapter 5 we examine nonlinear phenomena of the Korteweg-de Vries equation. We demonstrate that *Mathematica* is an appropriate tool to derive numerical and analytical solutions even for nonlinear equations of motion. Problems of general relativity are discussed in Chapter 6. Most standard books on Einstein's theory discuss the phenomena of general relativity by using approximations. With *Mathematica*, general relativity effects like the shift of the perihelion can be tracked with precision. Finally, the last chapter uses computer algebra to represent fractals and gives an introduction to the spatial renormalization theory. Exercises with which *Mathematica* packages can be used for modified applications are included at the end of Chapters 2–7.

The author would like to acknowledge the pre-publication reviewers of my manuscript, since they provided me with valuable suggestions for improving the materials: Paul Abbott, University of Western Australia; James Feagin, California State University, Fullerton; Alec Schramm, Occidental College, Los Angeles. An acknowledgement goes also to Patrick Tam, California State University, Humboldt, regarding the driven pendulum.

Ulm, Germany   *Gerd Baumann*
Fall 1995

Contents

1
Introduction

1.1 Basics

Mathematica is a computer algebra system which allows

- symbolic
- numeric
- graphical and
- acoustic

calculations. *Mathematica* was developed by Stephen Wolfram in the 1980s and is available for a large number of computers (PC, HP, SGI, SUN, NeXT, VAX, etc).

It is also possible to create custom programs by using *Mathematica*'s interactive definitions in a notebook. This capability allows us to solve physical problems directly on the computer. Before discussing the solution steps for several problems of theoretical physics, we present a short overview of the organization of *Mathematica*.

1.1.1 Structure of *Mathematica*

Mathematica and its parts consist of five main components (see figure 1.1):

- the *kernel*
- the *Frontend*
- the standard *Mathematica* packages
- the *MathSource* library
- the programs written by the user.

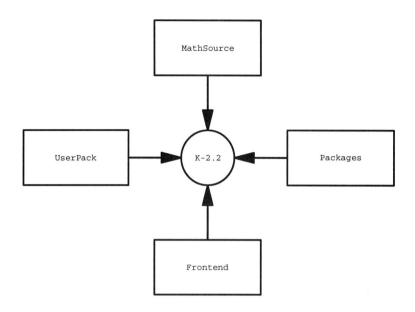

Figure 1.1. *Mathematica* system.

The *kernel* is the main engine of the system containing all the functions defined in *Mathematica*. The *Frontend* is the part of the *Mathematica* system serving as the channel on which a user communicates with the kernel. All components interact in a certain way with the kernel of *Mathematica*.

The kernel itself consists of about 1112 functions available after the initialization of *Mathematica*. The kernel manages calculations such as symbolic differentiations, symbolic integrations, graphical representations, evaluations of series and sums, etc.

The standard packages delivered with *Mathematica* contain a collection of special topics in mathematics. The contents of the packages range from vector analysis, statistics, algebra, to graphics, etc. A detailed description is contained in the technical report: Guide to Standard *Mathematica* Packages [1.4] published by Wolfram Research Inc.

MathSource is another source of *Mathematica* packages. *MathSource* consists of a collection of packages and notebooks created by *Mathematica* users for special purposes. For example, there are calculations of Feynman diagrams in high energy physics and Lie symmetries in the solution theory of partial differential equations. *MathSource* is available on the Internet via ftp by means of an anonymous account at *mathsource.wri.com*. It is also available on CD-ROM.

The last part of the *Mathematica* environment is created by each indivdual user. *Mathematica* allows each user to define new functions extending the functionality of *Mathematica* itself. This book belongs to this user-defined part of *Mathematica*.

The goal of our application of *Mathematica* is to show how problems of physics, mathematics and engineering can be solved. We use this computer program to support our calculations either in an interactive form or by creating packages which tackle the problem. We also show how non-standard problems can be solved using *Mathematica*.

However, before diving into the ocean of physical problems, we will first discuss some elementary properties of *Mathematica* that are useful for the solutions of our examples. In the following, we give a short overview of the capabilities of *Mathematica* in symbolic, numeric and graphical calculations. In a short section at the end of this chapter, the interactive use of *Mathematica* is discussed.

1.1.2 Symbolic calculations

By symbolic calculations we mean the manipulation of expressions using the rules of algebra and calculus. The following examples give a quick idea of how to use *Mathematica*. We will use some of the following functions in the remainder of this book.

A function consists of a name and several arguments enclosed in square brackets. The arguments are separated by commas. One function frequently used in the solution process is the function **Solve[ ]**. **Solve[ ]** needs two arguments: the equation to be solved and the variable for which the equation is solved. For each *Mathematica* function you will find a short description of its functionality and its purpose if you type the name of the function preceded by a question mark. For example, the description of **Solve[ ]** is:

```
?Solve

Solve[eqns, vars] attempts to solve an equation or
    set of equations for the variables vars. Any
    variable in eqns but not vars is regarded as a
    parameter. Solve[eqns] treats all variables
    encountered as vars above. Solve[eqns, vars,
    elims] attempts to solve the equations for vars,
    eliminating the variables elims.
```

The help facility of *Mathematica* (**?** or **??**) gives us a short description of any function contained in the *kernel*. For a detailed description of the functionality, the reader should consult the book by S. Wolfram *Mathematica*: A System for Doing Mathematics by Computer [1.1].

Let us start with an example using **Solve[ ]** and a quadratic equation in **t**:

```
Solve[t^2-t+a==0,t]

          1 - Sqrt[1 - 4 a]            1 + Sqrt[1 - 4 a]
    {{t -> -----------------}, {t -> -----------------}}
                  2                           2
```

The differentiation of a function for an independent variable is calculated by using the **D[ ]** function which is used for ordinary or partial differentiation:

```
D[Sin[t],t]

Cos[t]
```

The integration of a function is executed by

```
Integrate[t^a,t]

 1 + a
t
------
1 + a
```

The calculation of a limit follows by

```
Limit[Sin[t]/t,t->0]

1
```

The expansion of a function **f[t]** in a Taylor series around **t=0** up to third order is given by

```
Series[f[t],{t,0,3}]

                        2      (3)    3
                 f''[0] t    f   [0] t        4
 f[0] + f'[0] t + --------- + ---------- + O[t]
                     2            6
```

The calculation of a finite sum follows from

```
Sum[(1/2)^n,{n,1,10}]

1023
----
1024
```

The Laplace transform of the function **Sin[t]** is calculated with using the standard package **LaplaceTransform**, located in the subdirectory *Calculus* on your computer. **LaplaceTransform** is accessed by the **Get[ ]** function abbreviated by <<.

```
<<Calculus'LaplaceTransform'

LaplaceTransform[Sin[t],t,s]

   1
------
     2
1 + s
```

Ordinary differential equations can be solved using the function **DSolve[ ]**. A practical example is given by the relaxation equation $u' + \alpha u = 0$. The solution of this equation follows from:

```
DSolve[D[u1[t],t]+a u1[t] ==0,u1[t],t]

            C[1]
{{u1[t] -> ----}}
            a t
           E
```

The calculus of vector analysis can be treated using the package **VectorAnalysis** which contains useful functions for using cross products of vectors as well as for calculating gradients of scalar functions. Some examples of this kind of calculation follow:

```
<<Calculus'VectorAnalysis'

CrossProduct[{a,b,c},{a1,b1,c1}]

{-(b1 c) + b c1, a1 c - a c1, -(a1 b) + a b1}

Grad[f[x,y,z]]

   (1,0,0)              (0,1,0)
{f        [x, y, z], f        [x, y, z],
   (0,0,1)
  f        [x, y, z]}
```

This collection of symbolic calculations gives an idea of how the capabilities of *Mathematica* support our everyday calculations. Besides the symbolic calculations, we sometimes need the numerical evaluations of some expressions. Some examples for numerical calculations are given below.

1.1.3 Numerical calculations

The numerical capabilities of *Mathematia* allow the following three essential operations for solving practical problems.

The solution of equations; e.g., the solution of the 6th order polynomial $x^6 + x^2 - 1 = 0$ follows by

```
NSolve[x^6+x^2-1==0,x]

{{x -> -0.826031}, {x -> -0.659334 - 0.880844 I},
 {x -> -0.659334 + 0.880844 I},
 {x -> 0.659334 - 0.880844 I},
 {x -> 0.659334 + 0.880844 I}, {x -> 0.826031}}
```

If you need to evaluate a definite integral in the range $x \in [0, \infty]$ you can use the numerical integration capabilities of **NIntegrate[ ]**. An example from statistical physics is

```
NIntegrate[x^3 Exp[-x^4],{x,0,Infinity}]

0.25
```

Sometimes it is hard to find an analytic solution of an ordinary differential equation (ODE). The problem becomes much worse if you try to solve a nonlinear ODE. If you face such a situation in your work the function **NDSolve[ ]** may help you to tackle the problem. An example for a second order nonlinear ODE used in the examination of nonlinear oscillators is used to demonstrate the solution of the initial value problem $y'' - y^2 + 2y = 0$, $y(0) = 0$, $y'(0) = 1/2$. The initial value problem describes a nonlinear oscillator starting at $t = 0$ with a vanishing elongation and an initial velocity of $1/2$. The formulation in *Mathematica* reads

```
NDSolve[{D[y[t],{t,2}] - y[t]^2 + 2 y[t]==0,
         y[0]==0,y'[0]==1/2},y[t],{t,0,10}]

{{y[t] -> InterpolatingFunction[{0., 10.}, <>][t]}}
```

The result of the numerical integration is a representation of the solution by means of an interpolating function.

The above three examples may serve to demonstrate that *Mathematica* is also capable of handling numerical evaluations. There are a lot of other functions which support numerical calculations. As a rule, all functions which are involved with numerical calculations start with a capital **N** in the name. As mentioned at the beginning of this chapter the third topic in *Mathematica* is the graphical representation of the calculations.

1.1.4 Graphics

Mathematica supports the graphical representation of different mathematical expressions. *Mathematica* is able to create two- and three-dimensional plots. It allows the representation of experimental data given by lists of points, by parametric plots for functions in parametric form, or by contour plots for three-dimensional functions. It further allows the creation of short motion pictures by its function **Animation**. An overview of these capabilities is given below.

As a first example of the graphical capabilities of *Mathematica* let us show how simple functions are plotted. The first argument of the plot function **Plot[ ]** specifies the function, the second argument denotes the plot range. All other arguments are options which alter the form of the plot in some way. A standard example (see figure 1.2) frequently encountered is

```
Plot[Sin[x],{x,-Pi,Pi},AxesLabel->{"x","Sin[x]"}]

-Graphics-
```

In three dimensions we use **Plot3D[ ]** to represent the surface of a function. An example showing the surface in a rectangular water tank is given below. The arguments of **Plot3D[ ]** are similar to the function **Plot[ ]**. The first specifies the function, the second and third the plot range. All others are optional. The result is shown in figure 1.3.

```
Plot3D[Sin[x] Cos[y],{x,-Pi,Pi},{y,-2Pi,2Pi},
       AxesLabel->{"x","y","z"},
       PlotPoints->25]

-SurfaceGraphics-
```

Sometime you find that you know a solution only in a parametric representation. For example, if you examine the motion of an electron in a constant magnetic field, the track of the electron is described by a three-dimensional vector depending parametrically on time t. To represent such a parametric path you can use the

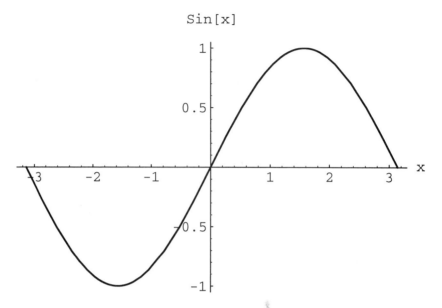

Figure 1.2.   Two-dimensional graphics.

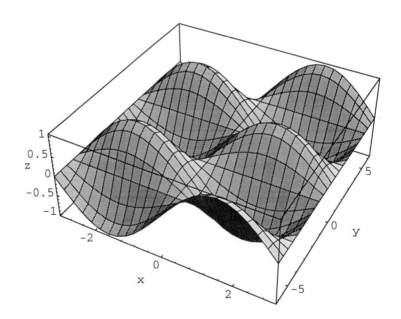

Figure 1.3.   Three-dimensional graphics.

function **ParametricPlot3D[ ]**. The first argument of this function contains a list which describes the three coordinates of the curve. A fourth element of this list (which is optional) allows to set a color for the track. In the following example we used the color function **Hue[ ]**. The second argument of the function **ParametricPlot3D[ ]** specifies the plot range of the parameter. All other arguments given to **ParametricPlot3D[ ]** are options changing the appearance of the plot (see figure 1.4).

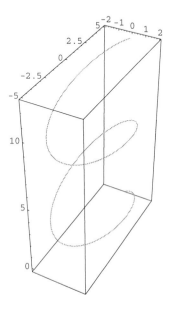

Figure 1.4.   Parametric plot.

```
ParametricPlot3D[{2Sin[t],5Cos[t],t,Hue[0.4]},
{t,0,4Pi}]

-Graphics3D-
```

In other situations like the movement of a planet around the sun you know the solution of the problem in an implicit form. From Keplers theory you know that a planet moves on an elliptic track around the sun (see figure 1.5). The path of the planet is described by a formula like $x^2 + 2y^2 = 3$. To graphically represent such a path, we can use a function known as **ImplicitPlot[ ]** in *Mathematica*. This function becomes available if we load the standard package **Graphics'ImplicitPlot'**. A representation of the hypothetical planet track in x and y follows for the range $x \in [-2, 2]$

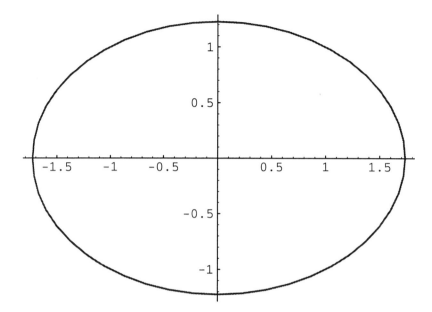

Figure 1.5.   Implicit plot.

```
<<Graphics'ImplicitPlot'

ImplicitPlot[x^2+2y^2==3,{x,-2,2},
PlotStyle->RGBColor[1,0,0]]

-Graphics-
```

The color of the curve is changed from black to red by the option **PlotStyle−>-RGBColor[1,0,0]**. If you have a function which is defined over a large range in x and y it is sometimes useful to represent the function in a log–log plot. For example, to show the graph of a scaling function like $f(x) = x^{1.4}$ in the range $x \in [1, 10^3]$, we can use **LogLogPlot[]** from the standard package **Graphics'Graphics'** to show the scaling behavior of the function (see figure 1.6). We clearly observe in the double logarithmic representation a linear relation between y and x which is characteristic for scaling (compare for example the theory of critical exponents in phase transitions).

```
<<Graphics'Graphics'

LogLogPlot[x^1.4,{x,1,1000},AxesLabel->{"x","y"}]

-Graphics-
```

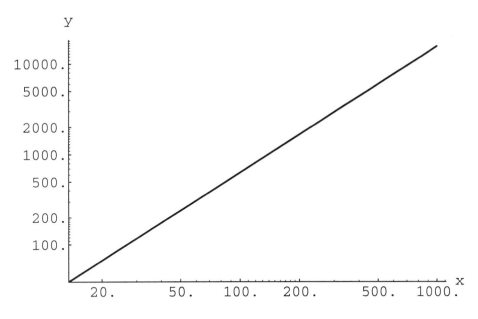

Figure 1.6.   Log-Log plot.

If you handle data from experiments *Mathematica* can do a lot of work for you. The graphical representation of a set of data can be done by the function **List-Plot**[ ]. This function allows you to plot a list of data. The input here is created by means of the function **Table**[ ]. The data set, which we will represent by **List-Plot**[ ] consists of pairs $\{x, \sin(x)\exp -x/4\}$ in the range $x \in [0, 6\pi]$. The data are located in the variable **tab1**. The graphical representation of these pairs of data is achieved by the function **ListPlot**[ ] using the data set **tab1** as first argument (see figure 1.7). All other arguments are used to temporarily set options for the function.

```
tab1 = Table[{x,Sin[x] Exp[-x/4]},{x,0,6Pi,.2}];

ListPlot[tab1,PlotStyle->RGBColor[0,0,1],
         AxesLabel->{"x","y"},
         PlotRange->All]

-Graphics-
```

If you need to represent several sets of data in the same figure, you can use the function **MultipleListPlot**[ ] contained in the standard package **Graphics'MultipleListPlot'**. An example for two sets of data **tab1** and **tab2** is given in figure 1.8 which we can create by

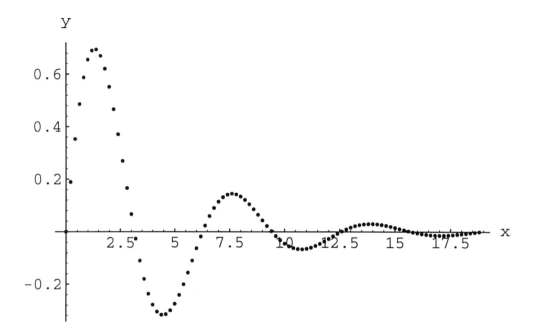

Figure 1.7.   Experimental data in a list plot.

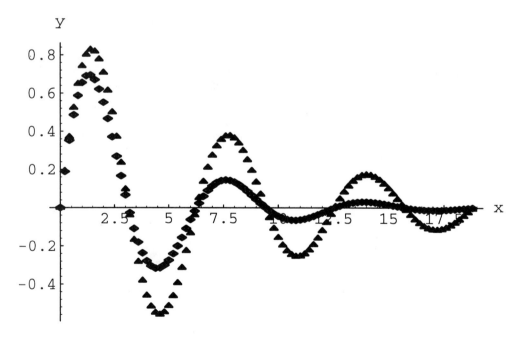

Figure 1.8.   Multiple list plot.

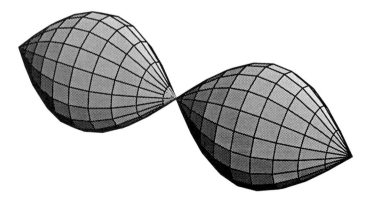

Figure 1.9.   Double bowl.

```
<<Graphics`MultipleListPlot`

tab2 = Table[{x,Sin[x] Exp[-x/8]},{x,0,6Pi,.2}];

MultipleListPlot[tab1,tab2,
        AxesLabel->{"x","y"},
        PlotRange->All]

-Graphics-
```

Sometimes results found by laborious calculations are poorly represented by simple pictures and there might be a way to "dress them up" a bit. In many situations you can vary a parameter or simply the time period to change the result in some way. The output of a small variation in parameters can be a great number of frames which all show different situations. To collect all the different frames in a common picture you can use the animation facilities of *Mathematica*. The needed functions are accessible if we load the standard package **Graphics'Animation'**. By using the functions contained in this package you can create, for example, a spinning bowl from a double bowl shown in figure 1.9. The spinning bowl is created by the function **SpinShow**[ ]. Different frames of the spinning bowl are shown in figure 1.10.

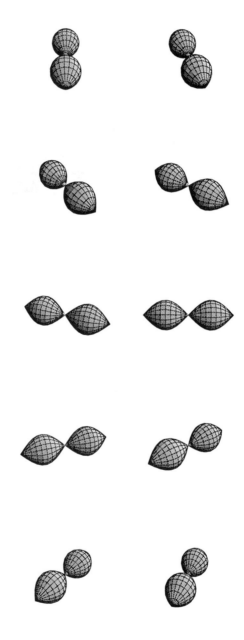

Figure 1.10.   Animation of the double bowl represented in a graphical array.

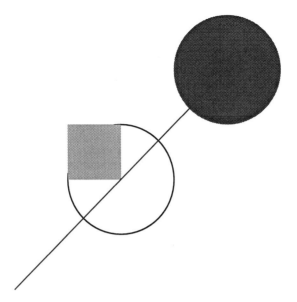

Figure 1.11.   Two-dimensional graphics primitives.

```
<<Graphics`Animation`

pl1 = ParametricPlot3D[{x,Cos[t] Sin[x],Sin[t] Sin[x]},
      {x,-Pi,Pi},{t,0,2Pi},
      Axes->False,Boxed->False]

-Graphics3D-
```

```
pl2 = SpinShow[pl1,Frames->10,SpinRange->{0 Degree, 180 Degree}];

Show[GraphicsArray[Partition[pl2,2]]]

-GraphicsArray-
```

If *Mathematica* does not provide you with the graphics you want, you are free to create your own graphics objects. By using graphics primitives like **Line[ ]**, **Disk[ ]**, **Circle[ ]** etc. you can create any two- or three-dimensional object you can imagine (figure 1.11). A simple example to combine lines, disks, squares, and circles follows.

```
l1 = {Circle[{0,0},1],{Hue[0.6],Disk[{2,2},1]},
      {Hue[0.3],Rectangle[{-1,1},{0,0}]},
      {Hue[0],Line[{{-2,-2},{2,2}}]}};

Show[Graphics[l1],AspectRatio->Automatic]

-Graphics-
```

1.1.5 Interactive use of *Mathematica*

Mathematica employs a very simple and logical syntax. All functions are accessible by their full names. The first letter of each name is capitalized. For example, if we wish to terminate our calculations and exit the *Mathematica* environment, we type the termination function **Quit[ ]**. Any function under *Mathematica* can be accessed by its name followed by a pair of square brackets which contain the arguments of the respective function. An example would be **Plot[Sin[x],{x,0,Pi}]**. The termination function **Quit[ ]** is one of the few functions that lacks an argument.

After activating *Mathematica* on the computer by typing *math* for the interactive version or *mathematica* for the notebook version, we can immediately go to work. Let us assume we need to calculate the ratio of two integer numbers. To get the result, we simply type in the expression and press return in the interactive or shift plus return in the notebook version. The result is a simplified expression of the rational number.

```
In[1]:= 69/15

          23
Out[1]= --
          5
```

The input and output lines of *Mathematica* carry labels counting the number of inputs and outputs in a session. The input label is **In[no]:=** while the related output label is **Out[no]=**. Another example is the exponentiation of a number. Type in

```
In[2]:= 2^10
```

and you will get

```
Out[2] = 1024
```

Multiplication of two numbers can be done in two ways. In this book the multiplication sign is replaced by a blank:

```
In[3]:= 2 5

Out[3]= 10
```

You can also use a star to denote multiplication:

```
In[4]:= 2*5

Out[4]= 10
```

Besides basic operations such as addition (+), multiplication (*), division (/), subtraction (-), and exponentiation (^), *Mathematica* knows a large number of mathematical functions, including the trigonometric functions **Sin**[] and **Cos**[], the hyperbolic functions **Cosh**[] and **Sinh**[], and many others. All available *Mathematica* functions are listed in the handbook by Stephen Wolfram [1.1]. Almost all functions listed in Abramowitz & Stegung [1.2] are also available in *Mathematica*.

By solving the following mathematical conjecture, we simultaneously demonstrate the creation of an interactive function in *Mathematica*. The iteration of the relation

$$f_{n+1} = f_{n-1} \int \left(\frac{f_n}{f_{n-1}} \right)^2 dx \tag{1.1}$$

under the initial conditions **f0** = **Cos**[x] and **f1** = **Sin**[x] results in a polynomial whose coefficients are given by trigonometric functions. The resulting polynomial can be represented in the form

$$f_\infty = \cos(x) \sum_{n=0}^{\infty} a_n x^n + \sin(x) \sum_{n=0}^{\infty} b_n x^n. \tag{1.2}$$

The related *Mathematica* representation is located in the variable **poly**. It reads:

```
poly = Cos[x] Sum[a[n] x^n,{n,0,Infinity}] +
       Sin[x] Sum[b[n] x^n,{n,0,Infinity}]
```

The sums in the representation of the polynomial extend across the range $0 < n < \infty$ which is given in the second argument of the function **Sum[ ]**. In the first step of the calculation we introduce a list containing the initial conditions of the iteration. Lists in *Mathematica* are represented by a pair of braced brackets which contain the elements of the list separated by commas. To save the list for future use, we set the list equal to the variable **listf** by

```
listf = {Cos[x],Sin[x]}

{Cos[x], Sin[x]}
```

The first iteration step is executed by the sequence

```
AppendTo[listf,listf[[1]]
          Integrate[(listf[[2]]/listf[[1]])^2,x]]

{Cos[x], Sin[x], -(x Cos[x]) + Sin[x]}
```

in which we append the result from an integration of the iteration formula to **listf** by means of the function **AppendTo[ ]**. The next step of the iteration takes the form

```
AppendTo[listf,listf[[2]]
          Integrate[(listf[[3]]/listf[[2]])^2,x]]

{Cos[x], Sin[x], -(x Cos[x]) + Sin[x],
     2                         3
  -3 x  Cos[x] + 3 x Sin[x] - x  Sin[x]
  -------------------------------------}
                  3
```

in which we increase the indices of the list elements in **listf** by one. The next interactive step results in

```
AppendTo[listf,listf[[3]]
          Integrate[(listf[[4]]/listf[[3]])^2,x]]

{Cos[x], Sin[x], -(x Cos[x]) + Sin[x],
     2                         3
  -3 x  Cos[x] + 3 x Sin[x] - x  Sin[x]
  -------------------------------------,
                  3
```

```
((-(x Cos[x]) + Sin[x])
        4              6              3
    (15 x  Cos[x] - x  Cos[x] - 15 x  Sin[x] +
        5
      6 x  Sin[x])) / (45 (x Cos[x] - Sin[x])))}
```

At this stage the resulting list is not in its simplest form. We simplify **listf** with function **Simplify[ ]**. The result of **Simplify** is stored into the variable **listf** which means that the old value of **listf** is overwritten by the new value resulting from the simplification.

```
listf = Simplify[listf]

{Cos[x], Sin[x], -(x Cos[x]) + Sin[x],
                                  3
        2                        x  Sin[x]
  -(x  Cos[x]) + x Sin[x] - ---------,
                                  3
    3                3
  (x  (-15 x Cos[x] + x  Cos[x] + 15 Sin[x] -
            2
        6 x  Sin[x])) / 45}
```

Applying the function **Plus[ ]** to **listf** adds all elements of the list together, resulting in the representation of the polynomial in the form

```
poly = Apply[Plus,listf]

                      2
Cos[x] - x Cos[x] - x  Cos[x] + 2 Sin[x] +
                3
              x  Sin[x]
  x Sin[x] - --------- +
                3
    3                3
  (x  (-15 x Cos[x] + x  Cos[x] + 15 Sin[x] -
            2
        6 x  Sin[x])) / 45
```

The coefficients of the trigonometric functions **Cos[ ]** and **Sin[ ]** are accessed by

```
Coefficient[poly,Cos[x]]

          2
  1 - x - x
```

```
Coefficient[poly,Sin[x]]

          3
          x
2 + x - --
          3
```

verifying the conjecture that the resulting function of the iteration is a polynomial
with coefficients **Cos[x]** and **Sin[x]**. The disadvantage of this calculation is that we
need to repeat the iteration. To avoid such repetition, we define a function which
performs the repetition automatically. Function **Iterate[ ]** derives the polynomial
up to an iteration order n.

```
Iterate[initial_List,maxn_]:=Module[
(* --- local variables --- *)
   {df={},dfh},
(* --- set f to the initial values --- *)
   f = initial;
(* --- iterate the formula and collect the results --- *)
   Do[AppendTo[f,
      f[[n]] Integrate[(f[[n+1]]/f[[n]])^2,x]],
   {n,1,maxn}];
(* --- calculate the sum of all elements in f --- *)
   f = Expand[Apply[Plus,Simplify[f]]];
(* --- extract the coefficients from the polynom --- *)
   fh = {Coefficient[f,initial[[1]]]};
   AppendTo[fh,Coefficient[f,initial[[2]]]];
(* --- return the result --- *)
   fh
   ]
```

```
Iterate[{Cos[x],Sin[x]},4]

             4     6     7        9
       2    x     x     x     2 x
{1 - x - x  - -- + -- - -- + ----,
             3    45    45    945
            5     6     8      10
       2 x    x     x      x
2 + x - ---- + -- - --- + ----}
        15     45   105   4725
```

The result is a list containing the polynomial coefficients of the **Cos[ ]** and **Sin[ ]**
functions. With **Iterate[ ]** we can change the conjecture in the following way. What
happens if we use as initial conditions hyperbolic functions instead of trigonometric
functions? The result is easy to derive if we use **Iterate[ ]** in the form of

```
Iterate[{Cosh[x],Sinh[x]},3]

                    4     6           5
            2     x     x         2 x
   {1  +  x  -  x   +  --  +  --,  x  -  ----}
                    3     45          15
```

Again we obtain a polynomial whose coefficients are given by hyperbolic functions.

This small example demonstrates the capabilities of *Mathematica* for finding solutions to a specific problem allowing us at the same time to modify the initial question. However, the iterative solution of the conjecture is not an exact proof. It only demonstrates the correctness of the conjecture empirically. Yet, the empirical proof of the conjectured behavior is the first step in proving the final result.

From the above example we have seen that the use of *Mathematica* facilitates our work insofar as special functions become immediately available to us, not only analytically, but also numerically and graphically. This notwithstanding, we first need to be able to understand the physical and mathematical relationships before we can effectively use *Mathematica* as a powerful tool.

In the following chapters we will demonstrate how problems occurring in theoretical physics can be solved by the use of *Mathematica*. Note that we will not provide the reader with a detailed description of *Mathematica*. Instead we will present a collection of mathematical steps gathered in a package. This package is useful for solving specific physical or mathematical problems by applying *Mathematica* as a tool. For a detailed description of the *Mathematica* functions we refer the reader to the handbook by Stephen Wolfram, *Mathematica*: A System for Doing Mathematics by Computer [1.1] or *Mathematica*: A Practical Approach by Nancy Blachman [1.3]. However, we hope that the reader will readily understand the solutions, since the code corresponds to notations in theoretical physics.

2

Classical Mechanics

2.1 The mechanics of discrete systems

Classical mechanics is one of the cornerstones of theoretical physics. The fundamental concepts of classical mechanics have not only a far reaching influence on field theory but also on quantum mechanics. One of its basic concepts is the formulation of equations of motion by means of a Poisson bracket. In order to determine such equations, the structure of the Poisson bracket as well as a generating function need to be known. In most cases, the generating function is the Hamiltonian H. If the coordinates by which H is expressible are canonical coordinates, then their related Poisson bracket is defined by

$$\{A, B\} = \sum_{n=1}^{N} \left(\frac{\partial A}{\partial q_n} \frac{\partial B}{\partial p_n} - \frac{\partial A}{\partial p_n} \frac{\partial B}{\partial q_n} \right). \tag{2.1}$$

Here q_n and p_n denote the $2N$ canonical coordinates. Once we know the Hamilton function, we are able to determine the equations of motion by means of the Poisson bracket

$$\dot{q}_m = \{q_m, H\} \tag{2.2}$$
$$\dot{p}_m = \{p_m, H\}. \tag{2.3}$$

As a simple example let us consider the harmonic oscillator. We know that the Hamiltonian is given by a quadratic function in the coordinate q and momentum p

$$H = \frac{p^2}{2m} + \frac{k^2}{2} q^2. \tag{2.4}$$

The Poisson bracket pertaining to a two dimensional phase space is of the form

$$\{A, B\} = \frac{\partial A}{\partial q} \frac{\partial B}{\partial p} - \frac{\partial A}{\partial p} \frac{\partial B}{\partial q}. \tag{2.5}$$

Setting out from these basic principles, we represent these relationships in *Mathematica* almost the same way as written with paper and pencil:

```
In[1]:= H = p^2/(2m) + k^2 q^2/2
               p^2      k^2
Out[1]=        ---   +  ---- q^2
               2m        2
```

After executing this input, we know that the expression $p^2/(2m) + k^2q^2/2$ is contained in the variable H. The equal sign $=$ assigns an expression from the right hand side of the equation to a variable on the left hand side. This assignment defines a transformation rule which applies at all times in *Mathematica*. **H** now contains this value until a new expression is defined for **H**. If we consider the Poisson bracket as an operator which is applied to two functions $A[q, p]$ and $B[q, p]$, then we get a new function in the coordinates p and q. Depending on the canonical variables, the Poisson bracket can be described in *Mathematica* as follows:

```
In[2]:= PoissonBracket[a_,b_]:= D[a,q] D[b,p]-D[a,p] D[b,q]
```

To define the **PoissonBracket**[ ], we apply a delayed assignment using the definition sign :=. The function is therefore only executed upon calling it by name. In this case, we identify the operation of differentiation. Note that the two arguments **a** and **b** are used in conjunction with a blank. This means that **a** and **b** may be replaced by any expression on the right hand side of our definition. **a_** and **b_** act as dummy variables for any kind of expression. The right hand side of our definition consists of the actual operational procedure, which in this case is the difference between two sets of products of two derivatives. The built–in function **D[a,q]** calculates the derivative of the expression **a** with respect to **q**. We can calculate the right hand side of the equations of motion (2.2–2.3) with the following commands:

```
In[3]:= PoissonBracket[q,H]
             p
Out[3]=    -
             m
In[4]:= PoissonBracket[p,H]
Out[4]=  -(k^2 q)
```

The disadvantage of the function **PoissonBracket**[ ] is that **p** and **q** are fixed variables inside the bracket so that **H** must be defined in the coordinates **p** and **q**. Our calculation however is done in such a way that there is no difference between the internal and external use of **p** and **q**. **p** and **q** are known within and outside the function **PoissonBracket[]**. Mistakes in our calculation may occur at this point because of the global use of variables involved. For example, if somewhere in our calculation we put q equal to the variable **x**, **q=x**, we would no longer differentiate with respect to **q** but with respect to **x**. To avoid such inconsistency, we need to define the variables **p** and **q** as local variables. This is done by introducing the **Block[]** environment. Another minor disadvantage of the **PoissonBracket** is that this function is only defined for a two-dimensional phase space. To eliminate this disadvantage, we extend the definition of the **PoissonBracket** by introducing N pairs of variables $(p[i], q[i])$ in the following form:

```
PoissonBracket[a_, b_, n_] :=
  Block[{pk}, pk =
    Simplify[Sum[D[a, q[i]] D[b, p[i]] - D[a, p[i]] D[b, q[i]], {i, 1, n}]]]
```

The *Mathematica* function **Block[]** allows you to set up an environment in which the values of variables can temporarily be changed. The same behavior results by using the function **Module[]**. When you execute a block, values assigned to the first argument are cleared. This mechanism allows you to decouple the calculations in a block from the rest of the *Mathematica* environment. This means that the variable **pk** only exists within the function **PoissonBracket[]** while **p[i]** and **q[i]** coordinates are still global expressions. The list {**pk**} containing the variable **pk** defines the local variables of the **PoissonBracket[]** within the **Block[]**.

The command **Simplify[]** simplifies the end result of the summation done by **Sum[]**. The Hamiltonian is now expressed in the coordinates **p[1]**,**p[2]**,... and **q[1]**,**q[2]**,.... The numbers in brackets assume the meaning of indices with respect to the variables q_i, p_i $(i = 1, 2, \ldots)$. For a two dimensional system with anharmonic coupling, for example, the Hamiltonian is

```
H = (p[1]^2 + p[2]^2)/2 + q[1]^2 q[2]^2 + (q[1]^2 + q[2]^2)/2
```

where the system parameters such as mass m and spring constant k have been set to unity $(m = k = 1)$.

When applying the **PoissonBracket[]** to the Hamiltonian for the coordinates q_1, q_2 and p_1, p_2, we get

```
In[1]:= PoissonBracket[q[1],H,2]

Out[1]= p[1]

In[2]:= PoissonBracket[q[2],H,2]
```

```
Out[2]= p[2]

In[3]:= PoissonBracket[p[1],H,2]

                    2
Out[3]=  -(q[1] (1 + 2 q[2] ))

In[4]:= PoissonBracket[p[2],H,2]

                   2
Out[4]= -((1 + 2 q[1] ) q[2])
```

A yet unsolved problem in our definition of the **PoissonBracket**[ ] is the choice of coordinates inside the bracket. So far we have used variables in the Poisson bracket which are predefined by **p[i]** and **q[i]**. Thus, the formulation of the Hamiltonian is always defined by the coordinates in the Poisson bracket. This disadvantage can be avoided by calling the **PoissonBracket**[ ] with additional arguments. A redefinition of our function **PoissonBracket**[ ] is given below.

The **If**[ ] function in the code below examines whether both lists **q** and **p** are of equal length. If this is so, we may proceed to calculate the Poisson bracket. If **q** and **p** are not of equal length, a message appears. The **If**[ ] function has three arguments. The first argument contains the conditions under which the next two arguments will be executed. If the first condition is true, then the second argument will be executed. If the first condition is false, then the third argument will be executed.

```
PoissonBracket[a_, b_, q_List, p_List]  :=
  Block[{pk,n},
    n = Length[q];
    If[n == Length[p],
      pk =
      Simplify[Sum[D[a, q[[i]]] D[b, p[[i]]] -
                  D[a, p[[i]]] D[b, q[[i]]], {i, 1, n}]],
        Print[" "];
        Print["  Length of lists does not agree! "]
      ]
      ]
```

Variables **q** and **p** are assigned to a specific variable type called **List**. Each function call examines whether **p** and **q** are lists. First, the function **PoissonBracket**[ ] examines whether both lists are of equal length. If so, the Poisson bracket is calculated; if not, an error message appears. Note that the variables according to which **q** and **p** are differentiated now possess two square brackets. This means that the i-th element of lists **q** and **p** are used as differentiation variables.

A definition of the function **PoissonBracket**[ ] within a given context environment allows a complete decoupling from the global *Mathematica* environment [2.1]. A context in *Mathematica* starts with the function **BeginPackage**[ ] which is assigned the context name and is terminated using **EndPackage[ ]**. The context environment in *Mathematica* can be compared to a directory tree in an operating system such as DOS or UNIX. It is the foundation for the design of *Mathematica* packages. The structure of our package for the Poisson bracket in the **Poisson'** environment is as follows:

```
────────────────────── File poisson.m ──────────────────────

BeginPackage["Poisson'"]

Clear[PoissonBracket]

PoissonBracket::usage = "PoissonBracket[a, b, q_List, p_List] calculates
the Poisson bracket for two functions a and b which depend on the variables
p and q. Example: PoissonBracket[q,p,{q},{p}] calculates the fundamental
bracket relation between the coordinate and momentum."

Begin["'Private'"]

(* --- definition of the Poisson bracket --- *)

PoissonBracket[a_, b_, q_List, p_List] :=
  Block[{pk,n},
(* --- check the length of the lists p and q --- *)
    n = Length[q];
    If[n == Length[p],
(* --- calculate the Poisson bracket   --- *)
      pk =
      Simplify[Sum[D[a, q[[i]]]*D[b, p[[i]]] -
                 D[a, p[[i]]]*D[b, q[[i]]], {i, 1, n}]],
(* --- display warnings --- *)
        Print[" "];
        Print["  length of lists does not agree! "]
      ]
       ]
End[]
EndPackage[]    (* --- no arguments needed --- *)
```

In the context environment **Poisson'**, our program package is defined anywhere between **BeginPackage[ ]** and **EndPackage[ ]** and may consist of several **Private** environments. All functions introduced as **::usage=** at the beginning of a package are general functions and are globally available in *Mathematica*. For a detailed description of these package techniques, the reader might refer to the book by R. Maeder [2.1].

In the next section we apply the Poisson bracket formulation to derive the equations of motion for a system consisting of two coupled harmonic oscillators.

2.2 Two coupled harmonic oscillators

The simplest case for coupled oscillators is that of a linear chain with two equal mass points $m_1 = m_2 = M$. The physical situation is shown in figure 2.1. To the left and right the mass points are attached to immovable walls by springs with spring constant k. In the center, there is one spring with spring constant k_{12}. The Hamiltonian of this linear chain is the sum of the potential and kinetic energies. Denoting the horizontal amplitudes of the mass points by x_1 and x_2, the total energy is given in the form of

$$H = \frac{1}{2M}(p_1^2 + p_2^2) + \frac{k}{2}(x_1^2 + x_2^2) + \frac{k_{12}}{2}(x_1 - x_2)^2. \qquad (2.6)$$

In *Mathematica*, the Hamiltonian takes the form of

```
In[1]:= H = (p1^2 + p2^2)/(2 M) + k (x1^2 + x2^2)/2 + k12 (x1-x2)^2/2
            2    2                    2    2     2

         p1  + p2    k12 (x1 - x2)    k (x1  + x2 )
Out[1]= --------- + -------------- + -------------
           2 M            2                2
```

Using **PoissonBracket[]** from the **Poisson'** package given in the previous section, we obtain the right hand sides of the Hamilton equations of motion. To use the **Poisson'** package, we first load the package by <<**Poisson.m**. By using **PoissonBracket[]**, we get a system of differential equations which are called Hamilton's equations.

```
In[2]:= p1t = PoissonBracket[p1,H,{x1,x2},{p1,p2}]
Out[2]= -(k x1) - k12 x1 + k12 x2
In[3]:= p2t = PoissonBracket[p2,H,{x1,x2},{p1,p2}]
Out[3]= k12 x1 - k x2 - k12 x2
In[4]:= x1t = PoissonBracket[x1,H,{x1,x2},{p1,p2}]
        p1
Out[4]= --
        M
In[5]:= x2t = PoissonBracket[x2,H,{x1,x2},{p1,p2}]
        p2
Out[5]= --
        M
```

Alternatively this system of four first order differential equations can be rewritten in the variables x_1 and x_2 as a system of two second order differential equations. By replacing the momentum $p_i = M\dot{x}_i \quad i = 1, 2$ and by assuming that x_1 and x_2 are given by harmonic solutions in the form of $x_1 = B_1 e^{i\omega t}$, $x_2 = B_2 e^{i\omega t}$, it follows that by the definitions below,

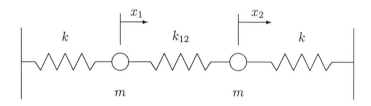

Figure 2.1.   Model of a linear chain.

```
In[6]:= x1 = B1 Exp[I omega t]
           I omega t
Out[6]= B1 E
In[7]:= x2 = B2 Exp[I omega t]
           I omega t
Out[7]= B1 E
```

an explicit representation of the solution is given in a harmonic form. This representation contains several unknowns which have to be determined in the course of our calculation.

With this formulation we can transform the system of second order differential equations into an algebraic system of equations which we will call **gl1** and **gl2**.

```
In[8]:= gl1 = Expand[(M D[x1, {t,2}] - p1t) Exp[-I omega t] ]
                                                       2
Out[8]= B1 k + B1 k12 - B2 k12 - B1 M omega
In[9]:= gl2 = Expand[(M D[x2, {t,2}] - p2t) Exp[-I omega t] ]
                                                       2
Out[9]= B2 k - B1 k12 + B2 k12 - B2 M omega
```

The unknown quantity in these equations is the frequency ω. In order to obtain a non–trivial solution for the two equations, the determinant of the coefficients of **B1** and **B2** must vanish. The matrix of coefficients is obtained as follows:

```
In[10]:= matrix = {{Coefficient[gl1,B1],Coefficient[gl1,B2]},
                   {Coefficient[gl2,B1],Coefficient[gl2,B2]}}
                           2                               2
Out[10]= {{k + k12 - M omega , -k12}, {-k12, k + k12 - M omega }}
```

Mathematica treats matrices as nested lists. A line–by–line input is stored in separate sublists. The accompanying determinant is determined with the function **Det[]**.

```
In[11]:= det = Det[matrix]
            2                    2              2   2      4
Out[11]= k  + 2 k k12 - 2 k M omega  - 2 k12 M omega  + M  omega
```

Since the determinant of non–trivial solutions must be zero, we have to solve a quartic equation for ω. Using **Solve[]**, we get

```
In[12]:= erg = Solve[det==0, omega]
                    Sqrt[k]              Sqrt[k]              Sqrt[k + 2 k12]
Out[12]= {{omega -> -------}, {omega -> -(-------)}, {omega -> ---------------},
                    Sqrt[M]              Sqrt[M]              Sqrt[M]
                    Sqrt[k + 2 k12]
    >    {omega -> -(---------------)}}}
                    Sqrt[M]
```

When applying **Solve[]**, we need to get an equation in the form of **left–hand side == right–hand side** as well as the accompanying unknown variable, separated by a comma. The solutions are comprised in a list identified as replacement rules indicated by $->$. Our results can be used in future calculations by using these replacement rules, as shown here.

Using ω_1 to denote the last two solutions and ω_2 to denote the first two solutions, the general solution for the oscillation problem takes the form

$$x_1(t) = B_{11}^+ e^{i\omega_1 t} + B_{11}^- e^{-i\omega_1 t} + B_{12}^+ e^{i\omega_2 t} + B_{12}^- e^{-i\omega_2 t} \tag{2.7}$$

$$x_2(t) = B_{21}^+ e^{i\omega_1 t} + B_{21}^- e^{-i\omega_1 t} + B_{22}^+ e^{i\omega_2 t} + B_{22}^- e^{-i\omega_2 t}. \tag{2.8}$$

Note that not all coefficients of these solutions depend on one another. We can verify this by inserting our solutions into the algebraic system of equations **gl1**, **gl2**. Calling

```
In[13]:= gl1h = gl1 /. erg[[3]]
Out[13]= B1 k + B1 k12 - B2 k12 - B1 (k + 2 k12)
In[14]:= gl2h = gl2 /. erg[[3]]
Out[14]= B2 k - B1 k12 + B2 k12 - B2 (k + 2 k12)
In[15]:= Solve[{gl1h==0,gl2h==0}, {B1,B2}]
Out[15]= {{B1 -> -B2}}
```

results in $B_{11} = -B_{21}$ for $\omega = \omega_1$.

We have taken the following steps: First we produced two auxiliary equations **gl1h** and **gl2h**, deviating from **gl1** and **gl2** whereby **omega** is replaced by the third element of the solution. The replacement was performed by the function **ReplaceAll[]** (abbreviated by /.). The rule of replacement is $omega->$ $\sqrt{k + 2k_{12}}/\sqrt{M}$; i.e., the third element of the list **erg**.

By calling **Solve[]** together with the auxiliary equation **gl1h** and **gl2h**, we obtain a list of solutions for the unknowns **B1** and **B2**. The unknown variables of the second argument of **Solve[]** are also included in a list. Again, the results are given by replacement rules.

The same procedure can be applied to the case of $\omega = \omega_2$. The result is

```
In[16]:= gl1h = gl1 /. erg[[1]]
Out[16]= B1 k12 - B2 k12
In[17]:= gl2h = gl2 /. erg[[1]]
Out[17]= -(B1 k12) + B2 k12
In[18]:= Solve[{gl1h==0,gl2h==0}, {B1,B2}]
Out[18]= {{B1 -> B2}}
```

The significance of the replacement rules $\{\mathbf{B1} -> \mathbf{B2}\}$ is that **B1** needs to be replaced by **B2** in equations **gl1h** and **gl2h**, or in the conventional mathematical notation $B_{12} = B_{22}$. The number of free constants in (2.7) is thus reduced to four, which is consistent with the number of integration constants of a coupled system of two second order equations. The general solution thus reads

$$x_1(t) = B_{11}e^{i\omega_1 t} + B_{12}e^{-i\omega_1 t} + B_{21}e^{i\omega_2 t} + B_{22}e^{-i\omega_2 t} \qquad (2.9)$$
$$x_2(t) = -B_{11}e^{i\omega_1 t} - B_{12}e^{-i\omega_1 t} + B_{21}e^{i\omega_2 t} + B_{22}e^{-i\omega_2 t} \qquad (2.10)$$

or in *Mathematica*

```
In[19]:= o1 = omega /. erg[[3]]
         Sqrt[k + 2 k12]
Out[19]= ----------------
             Sqrt[M]
In[20]:= o2  = omega /. erg[[1]]
         Sqrt[k]
Out[20]= -------
         Sqrt[M]
In[21]:= x1 = B11 Exp[I o1 t] +
             B12 Exp[-I o1 t] +
             B21 Exp[I o2 t] +
             B22 exp[-I o2 t]
             (-I Sqrt[k] t)/Sqrt[M]          (-I Sqrt[k + 2 k12] t)/Sqrt[M]
Out[21]= B22 E                       + B12 E                               +
                 (I Sqrt[k + 2 k12] t)/Sqrt[M]          (I Sqrt[k + 2 k12] t)/Sqrt[M]
       >   B11 E                               + B21 E
```

```
In[22]:= x2 = -B11 Exp[I o1 t] -
            B12 Exp[-I o1 t] +
            B21 Exp[I o2 t] +
            B22 exp[-I o2 t]
            (-I Sqrt[k] t)/Sqrt[M]          (I Sqrt[k] t)/Sqrt[M]
Out[22]= B22 E                          + B21 E                          -
                (-I Sqrt[k + 2 k12] t)/Sqrt[M]          (I Sqrt[k + 2 k12] t)/Sqrt[M]
         >    B12 E                                   - B11 E
```

Finally, the free constants $B_{11}, B_{12}, B_{21}, B_{22}$ are determined by the initial conditions $x_1(t = 0) = x_{10}$, $x_2(t = 0) = x_{20}$, $p_1(t = 0) = p_{10}$ and $p_2(t = 0) = p_{20}$. Since the momentum is proportional to the time derivative of the coordinates, the initial conditions are $(x_{10}, x_{20}, p_{10}, p_{20}) = (0, 0, 1, -1/2)$. We assume that x_{10} and x_{20} vanish and that the initial momenta p_{10} and p_{20} take finite values of $(1, 1/2)$ respectively. The initial conditions can be represented by

```
In[23]:= x10 = x1 /. t->0
Out[23]= B11 + B12 + B21 + B22
In[24]:= x20 = x2 /. t->0
Out[24]= -B11 - B12 + B21 + B22
In[25]:= p10 = D[x1,t] /. t->0
          -I B22 Sqrt[k]   I B11 Sqrt[k + 2 k12]   I B12 Sqrt[k + 2 k12]
Out[25]= -------------- + -------------------- - -------------------- +
             Sqrt[M]            Sqrt[M]                Sqrt[M]
              I B21 Sqrt[k + 2 k12]
         >    --------------------
                   Sqrt[M]
In[26]:= p20 = D[x2,t] /. t->0
          I B21 Sqrt[k]   I B22 Sqrt[k]   I B11 Sqrt[k + 2 k12]   I B12 Sqrt[k + 2 k12]
Out[26]= ------------- - ------------- - -------------------- + --------------------
             Sqrt[M]         Sqrt[M]           Sqrt[M]                Sqrt[M]
In[27]:= koerg = Solve[{x10==0,x20==0,p10==1,p20==-1/2},
                   {B11,B12,B21,B22}]
Out[27]= {{B11 -> (- (k    + 2 Sqrt[k] k12 - 7 k Sqrt[k + 2 k12] +
                   4
                                                           2            2
         >            k12 Sqrt[k + 2 k12]) Sqrt[M]) / (4 k  + 7 k k12 - 2 k12 ),
                     I   3/2
         >      B12 -> (- (k    + 2 Sqrt[k] k12 - 7 k Sqrt[k + 2 k12] +
                     4
                                                            2            2
         >             k12 Sqrt[k + 2 k12]) Sqrt[M]) / (-4 k  - 7 k k12 + 2 k12 ),
                     I
         >      B21 -> (- (3 Sqrt[k] - Sqrt[k + 2 k12]) Sqrt[M]) / (-4 k + k12),
                     4
                     -I
         >      B22 -> (-- (3 Sqrt[k] - Sqrt[k + 2 k12]) Sqrt[M]) / (-4 k + k12)}}
                     4
In[28]:= x1p = x1 /. koerg
          I  (-I Sqrt[k] t)/Sqrt[M]      3/2
Out[28]= {(- E                    (3 k    + 6 Sqrt[k] k12 - k Sqrt[k + 2 k12] -
          4
```

```
      >          2 k12 Sqrt[k + 2 k12]) Sqrt[M]) / (4 k  + 7 k k12 - 2 k12 ) +
                                                    2                2
                I  (-I Sqrt[k + 2 k12] t)/Sqrt[M]
      >        (- E
                4
                    3/2
      >          (-k    - 2 Sqrt[k] k12 + 7 k Sqrt[k + 2 k12] - k12 Sqrt[k + 2 k12])
                              2                2
      >          Sqrt[M]) / (4 k  + 7 k k12 - 2 k12 ) +
                I  (I Sqrt[k + 2 k12] t)/Sqrt[M]
      >        (- E
                4
                    3/2
      >          (k    + 2 Sqrt[k] k12 - 7 k Sqrt[k + 2 k12] + k12 Sqrt[k + 2 k12])
                              2                2
      >          Sqrt[M]) / (4 k  + 7 k k12 - 2 k12 ) +
                I  (I Sqrt[k + 2 k12] t)/Sqrt[M]
      >        (- E
                4
                    3/2
      >          (-3 k    - 6 Sqrt[k] k12 + k Sqrt[k + 2 k12] + 2 k12 Sqrt[k + 2 k12])
                              2                2
      >          Sqrt[M]) / (4 k  + 7 k k12 - 2 k12 )}
In[29]:= x2p = x2 /. koerg
                I  (-I Sqrt[k] t)/Sqrt[M]       3/2
Out[29]= {(- E                        (3 k    + 6 Sqrt[k] k12 - k Sqrt[k + 2 k12] -
                4
                                                      2                2
      >          2 k12 Sqrt[k + 2 k12]) Sqrt[M]) / (4 k  + 7 k k12 - 2 k12 ) -
                I  (-I Sqrt[k + 2 k12] t)/Sqrt[M]
      >        (- E
                4
                    3/2
      >          (-k    - 2 Sqrt[k] k12 + 7 k Sqrt[k + 2 k12] - k12 Sqrt[k + 2 k12])
                              2                2
      >          Sqrt[M]) / (4 k  + 7 k k12 - 2 k12 ) -
                I  (I Sqrt[k + 2 k12] t)/Sqrt[M]
      >        (- E
                4
                    3/2
      >          (k    + 2 Sqrt[k] k12 - 7 k Sqrt[k + 2 k12] + k12 Sqrt[k + 2 k12])
                              2                2
      >          Sqrt[M]) / (4 k  + 7 k k12 - 2 k12 ) +
                I  (I Sqrt[k] t)/Sqrt[M]       3/2
      >        (- E                        (-3 k    - 6 Sqrt[k] k12 + k Sqrt[k + 2 k12] +
                4
                                                      2                2
      >          2 k12 Sqrt[k + 2 k12]) Sqrt[M]) / (4 k  + 7 k k12 - 2 k12 )}
```

The variable **koerg** contains the solution for the coefficients **B11, B12, B21**
and **B22** for special initial conditions. The solutions for the variables **x1** and **x2**
require the **B** solutions in order to find an explicit expression for these special initial
conditions. This is done by applying the substitution rules   **koerg** to the general

representation of **x1** and **x2**. If we further specify the physical parameters such as mass and spring constants, we obtain the following solutions for $(M, k, k_{12}) = (1, 1, 2)$:

```
In[30]:= x1p = x1p /. {M->1,k->1,k12->2}
           I          3/2   -I t   I           3/2   -I Sqrt[5] t
Out[30]= {-- (15 - 5   ) E      + -- (-5 + 5   ) E                +
          40                      40

              I          3/2   I Sqrt[5] t   I            3/2   I Sqrt[5] t
          >   -- (5 - 5   ) E              + -- (-15 + 5   ) E              }
              40                             40
In[31]:= x2p = x2p /. {M->1,k->1,k12->2}
           I          3/2   -I t   I           3/2   I t   I           3/2   -I Sqrt[5] t
Out[31]= {-- (15 - 5   ) E      + -- (-15 + 5   ) E     - -- (-5 + 5   ) E                -
          40                      40                      40

              I          3/2   I Sqrt[5] t
          >   -- (5 - 5   ) E              }
              40
```

Only the real part of these solutions, illustrated in figure 2.2, is relevant. The figure is created by

Re[x1p],Re[x2p]

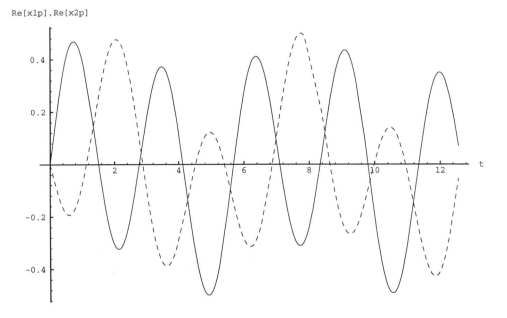

Figure 2.2.   The time history of the real part of the solutions of x_1 and x_2 (dotted) for $(M, k, k_{12}) = (1, 1, 2)$ with $(x10, x20, p10, p20) = (0, 0, 1, -1/2)$ as initial conditions.

```
Plot[{Re[x1p],Re[x2p]}, {x,0,50}]
```

This example demonstrates *Mathematica*'s interactive solution capabilities. For this procedure to succeed, however, we need to have a complete understanding of all agreed functions and definitions. In large projects involving a large number of variables and functions it is hard to keep track of the increasing number of defined variables. Thus we need an alternative approach to organize our work. An alternative method to interactive solutions is the development of small packages which allow a broad treatment of specific problems. In the following, our objective is to find solutions for problems with a minimum of interactive operations. This means that we first look at the theoretical concept of a solution which then serves as the basis for our programs. In section 2.4.2.1, we explain such a solution procedure by generalizing the two–particle problem to an N–member linear chain.

2.3 The pendulum

Solutions of certain nonlinear oscillation problems can be expressed in closed form in terms of elliptic integrals which are supported in *Mathematica*. The pendulum is one example of a nonlinear model exhibiting elliptic functions as solutions. A pendulum is a system with mass m which is kept in orbit by a massless supporting rod of length l (see figure 2.3). The pendulum moves within the gravitational field of the earth

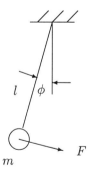

Figure 2.3.   The pendulum.

and is thus exposed to the vertical gravitational force mg. The dynamic force F is perpendicular to the supporting rod and takes the form $F(\phi) = -mg\sin(\phi)$.

For small amplitudes, we can model the pendulum in terms of a linear system which is equivalent to a harmonic oscillator. The accuracy of this approximation will be determined in the course of our calculations. The equation of motion can be derived from the equilibrium of the torque and the product of angular acceleration and the momentum of inertia with respect to the pivotal point:

$$I\ddot{\phi} = lF. \tag{2.11}$$

Since the moment of inertia of the pendulum is $I = ml^2$, we get an equation of motion in the form of

$$\ddot{\phi} + \omega_0^2 \sin(\phi) = 0 \tag{2.12}$$

with $\omega_0^2 = g/l$ being the ratio between the gravitational acceleration g and the length of the pendulum l. If the amplitudes around the equilibrium position are small, then $\sin(\phi)$ in (2.12) can be approximated by $\sin(\phi) \approx \phi$. As a result, the equation of motion is reduced to an equation of a harmonic oscillator

$$\ddot{\phi} + \omega_0^2 \phi = 0. \tag{2.13}$$

Within this approximation, the oscillation period T is given by $T = 2\pi/\omega_0 = 2\pi(l/g)^{1/2}$ and is independent of the amplitude.

If we wish to determine the oscillation period for larger amplitudes, we need to start with equation (2.12). Since we have neglected damping in our equations, the total energy of the system can be written as the sum of the potential and kinetic energy (conservation of energy)

$$T_{kin} + V = E = const. \tag{2.14}$$

This formulation allows us to easily construct the solution to (2.12). Equation (2.14) gives a first integral of motion. Due to the explicit time independence of the equation of motion (2.12) the second step of the integration process can be done by a quadrature. The duration of oscillation can now be expressed in the form of an integral.

If we choose the origin of the potential energy to be at the lowest point in the orbit, then we get for the potential energy

$$V = mgl(1 - \cos(\phi)). \tag{2.15}$$

The kinetic energy is derived from the equation

$$T_{kin} = \frac{1}{2}I\omega^2 = \frac{1}{2}ml^2\dot{\phi}^2. \tag{2.16}$$

If we designate the angle at the highest orbital point as ϕ_1, then the potential and the kinetic energy at this point are given by

$$V(\phi = \phi_1) \quad = \quad E = mgl(1 - \cos(\phi_1)) \tag{2.17}$$
$$T_{kin}(\phi = \phi_1) \quad = \quad 0. \tag{2.18}$$

By means of the trigonometric identity $\cos(\phi) = 1 - 2\sin^2(\phi/2)$, the total energy at the upper reversal point can be expressed in the form of

$$E = 2mgl\sin^2\left(\frac{\phi_1}{2}\right). \tag{2.19}$$

Since E is constant over time, this expression is also valid for amplitudes smaller than ϕ_1. The potential energy takes the form

$$V = 2mgl\sin^2\left(\frac{\phi}{2}\right) \tag{2.20}$$

by using the appropriate trigonometric identity $\cos(\phi) = 1 - 2\sin^2(\phi/2)$. In accordance with (2.14), the kinetic energy is given as the difference between the total energy and the potential energy by

$$\frac{1}{2}ml^2\dot{\phi}^2 = 2mgl\left\{\sin^2\left(\frac{\phi_1}{2}\right) - \sin^2\left(\frac{\phi}{2}\right)\right\}. \tag{2.21}$$

In other words, we get

$$\dot{\phi} = 2\omega_0\left\{\sin^2\left(\frac{\phi_1}{2}\right) - \sin^2\left(\frac{\phi}{2}\right)\right\}^{1/2}, \tag{2.22}$$

and separating the variables, we find

$$dt = \frac{1}{2\omega_0}\left\{\sin^2\left(\frac{\phi_1}{2}\right) - \sin^2\left(\frac{\phi}{2}\right)\right\}^{-1/2}d\phi. \tag{2.23}$$

We can obtain the oscillation period T of the pendulum by integrating both sides over a complete period

$$\int_0^T dt = 4\int_0^{\phi_1}\frac{1}{2\omega_0}\left\{\sin^2\left(\frac{\phi_1}{2}\right) - \sin^2\left(\frac{\phi}{2}\right)\right\}^{-1/2}d\phi. \tag{2.24}$$

The left hand side of (2.24) can be directly integrated and we find

$$T = \frac{2}{\omega_0}\int_0^{\phi_1}\frac{d\phi}{\sqrt{\sin^2\frac{\phi_1}{2} - \sin^2\frac{\phi}{2}}}. \tag{2.25}$$

Thus, the oscillation period is reduced to a complete elliptic integral. By substituting $z = \sin(\phi/2)/\sin(\phi_1/2)$ and $k = \sin(\phi_1/2)$, the integral on the right hand side of (2.25) is transformed to the standard form of

$$T = \frac{4}{\omega_0} \int_0^1 \frac{dz}{\sqrt{(1 - z^2)(1 - k^2 z^2)}} \tag{2.26}$$

$$= \frac{4}{\omega_0} K(k^2). \tag{2.27}$$

$K(k^2)$ denotes the complete elliptic integral of the first kind and $k^2 = El/(2g) = m$ denotes the modulus of the elliptic function.

By calling **EllipticK**[ ], *Mathematica* executes $K(k^2)$. **Integrate**[ ] executes the integration of (2.26):

```
4 Integrate[1/Sqrt[(1-z^2) (1-k^2 z^2)], {z,0,1}]/omega
```

and results in

```
4 EllipticK[k^2]/omega
```

Once we know the length of the pendulum and its initial angular displacement, the oscillation period is completely determined. Since *Mathematica* recognizes all elliptic integrals as well as all Jacobian elliptic functions, we can straightforwardly determine the dependence of the period on the initial amplitude. A graphical representation of $K(k)$ via ϕ_1 can be found in figure 2.4. We are now able to evaluate the period T with the following function:

```
T[omega_,phi1_]:=Block[{k,duration},
         k = Sin[phi1/2];
         duration = 4 EllipticK[k^2]/omega]
```

Our input values are the angle of displacement ϕ_1 and the frequency $\omega_0 = \sqrt{g/l}$. We first calculate the modulus k in accordance with the above definition and then determine the period in accordance with (2.27). As we see from figure 2.4, $K(k)$ with $k = 1$ tends toward ∞; i.e., at the upper point of reversal $\phi_1 = \pi$, the period is infinitely large.

Approximated equations are often cited in the literature [2.2] for the period. To obtain a valid comparison between exact and approximated oscillation periods, we use the approximation procedure described below. If the pendulum oscillates,

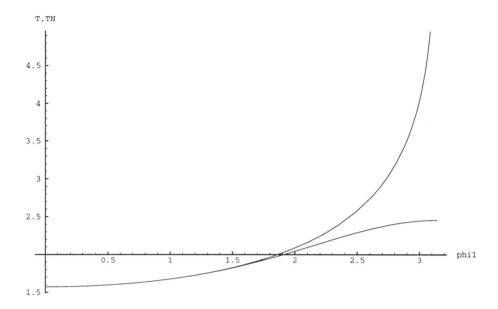

Figure 2.4. Comparison between the exact period T (upper curve) and the approximation T_N with an expansion up to the 8-th order with $\omega_0 = 4$.

we know that $k < 1$. Using this condition, we can expand the second part of the integrand in (2.26) into a Taylor series.

$$(1 - k^2 z^2)^{-1/2} = 1 + \frac{k^2 z^2}{2} + \frac{3k^4 z^4}{8} + \ldots . \qquad (2.28)$$

We execute this procedure using

```
res = Series[(1-k^2 z^2)^(-1/2), {k,0,8}]
```

The result is

```
      2  2      4  4      6  6      8  8
     z  k     3 z  k     5 z  k    35 z  k           9
1 + ----- + ------- + ------- + -------- + O[k]
       2        8        16        128
```

We have expanded the expression $(1 - k^2 z^2)^{-1/2}$ around $k = 0$ up to the 8th order by calling the function **Series[]** which yields a Taylor expansion. The period is expressed by using the Taylor representation

$$T_N = \frac{4}{\omega_0} \int_0^1 \frac{dz}{(1-z^2)^{1/2}} \left\{ 1 + \frac{k^2 z^2}{2} + \frac{3k^4 z^4}{8} + \ldots \right\}, \tag{2.29}$$

which in *Mathematica* looks as follows:

```
TN = 4 Integrate[Normal[res]/(1-z^2)^(1/2), {z,0,1}]/omega
```

By calling **Normal[ ]**, we eliminate the symbol $O[k]^9$ from the variable **res**. After executing the integration of the truncated expression **res** with **Integrate[ ]** and applying **Expand[ ]** to simplify the result, we get

```
          2        4         6            8
2 Pi    k  Pi   9 k  Pi   25 k  Pi   1225 k  Pi
----- + ------- + -------- + --------- + ----------
omega   2 omega  32 omega  128 omega   8192 omega
```

which agrees with the result given by Landau [2.2] with respect to the first set of orders

$$T_N \approx \frac{2\pi}{\omega_0} \left\{ 1 + \frac{k^2}{4} + \frac{9k^4}{64} + \ldots \right\}. \tag{2.30}$$

To use the same independent variables in a graphical representation, we replace k by $\sin(\phi_1/2)$. *Mathematica* executes such a replacement with the operator **ReplaceAll[ ]** (/.).

```
TN = TN /. k -> Sin[phi1/2]

                phi1 2          phi1 4            phi1 6
        Pi Sin[----]    9 Pi Sin[----]    25 Pi Sin[----]
2 Pi           2                2                  2
----- + ------------- + --------------- + --------------- +
omega      2 omega         32 omega          128 omega

                phi1 8
        1225 Pi Sin[----]
                    2
>       ------------------
           8192 omega
```

In order to get a graphical representation of this approximation we now need to specify a value for **omega** in **TN** to obtain an expression void of any parameter. To keep it simple, we choose **omega=4**. The replacement is executed by

```
TN = TN /. omega -> 4;
```

T and T_N can now be graphically presented as follows:

```
Plot[{T[4,phi1],TN}, {phi1,0,Pi},AxesLabel->{"phi1","T,TN"}]
```

Plot[ ] here is used together with a list of functions pertaining to the first argument. The second argument contains the range of representations. The third argument contains the axis labels.

Figure 2.4 shows that for small **phi1** the amplitudes between the exact period and its approximations are negligible. However, the difference between the exact theory and the approximation becomes larger and larger for angular displacements larger than $\phi_1 \approx 2$. In other words, for large ϕ_1, i.e., for large amplitudes, a larger number of higher order Taylor components is needed to obtain an accurate representation of the period.

If, however, we make the period dependent on the initial displacement ϕ_1 and note that k is connected to the initial condition via $k = \sin(\phi_1/2) \approx \phi_1/2 - \phi_1^3/48\ldots$, the range of agreement is further reduced by

$$T_N \approx \frac{2\pi}{\omega_0}\left\{1 + \frac{1}{16}\phi_1^2 + \frac{11}{3072}\phi_1^4 + \ldots\right\}. \tag{2.31}$$

The steps in *Mathematica* for this formulation are

```
sin = Series[Sin[phi1/2], {phi1,0,4}];
TN = TN  /. k->Normal[sin];
Expand[TN]
```

Series[ ] produces an expansion of **Sin** at $\phi_1 = 0$ up to the 4-th order. In the second step above, k in TN is replaced by the series expansion **sin** and is simplified by **Expand**[ ] in the last step. The result of this approximation is

```
          2           4             6              8
 2 Pi    phi1  Pi  11 phi1  Pi   25 phi1  Pi   9 phi1  Pi
 ----- + -------- + ----------- + ----------- + ------------- -
 omega   8 omega   1536 omega    73728 omega   2097152 omega
            10                 12              14                   16
       757 phi1   Pi   21773 phi1    Pi   8075 phi1    Pi    4675 phi1    Pi
  >   ------------- + --------------- - ---------------- + ------------------ -
       6291456 omega   905969664 omega   3623878656 omega   38654705664 omega
                 18                    20                      22
        25625 phi1    Pi        8575 phi1    Pi       1225 phi1    Pi
  >   ------------------- + -------------------- - -------------------- +
       6262062317568  omega   100192997081088  omega   1202315964973056  omega
                  24
          1225 phi1   Pi
  >   -------------------------
       230844665274826752  omega
```

Despite the limited accuracy, we can see from this approximation procedure that the period of a nonlinear problem depends on the initial conditions. In a borderline case of linear approximation, however, the period is independent of initial conditions.

2.3.1 Solutions for different values of energy

When we look at the potential $V(x) = 1 - \cos(x)$ for the pendulum, we observe by specifying the total energy E of the pendulum that three forms of motion are possible. For a total energy smaller than the maximum value of the potential energy, oscillations occur (bound motion). For energy values of $E > V_{max}$ we get rotations (see figure 2.5). Finally, for $E = V_{max}$ we get the asymptotic behavior of the pendulum. The solutions for the different values of energy result from (2.22) in the form of

$$\dot{\phi} = \pm\sqrt{\frac{2}{ml^2}\left\{E - ml^2\omega_0^2(1 - \cos(\phi))\right\}}. \tag{2.32}$$

Scaling the energy with $E^* = E/(ml^2\omega_0^2)$, we get

$$\dot{\phi} = \pm\omega_0\sqrt{2(\cos(\phi) - 1 + E^*)}. \tag{2.33}$$

Different forms of motion occur for different values of the scaled energy

$$E^* > 2 \qquad \text{rotation}$$
$$E^* = 2 \qquad \text{asymptotic motion}$$
$$0 \le E^* < 2 \qquad \text{oscillations.}$$

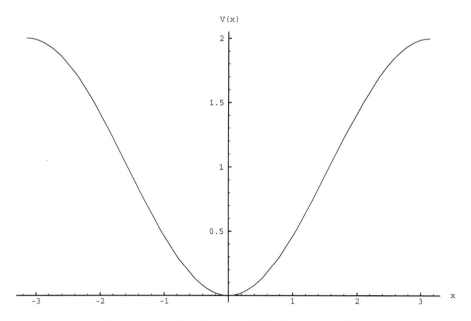

Figure 2.5. Scaled potential $V(x)$ for the pendulum.

In the following we will investigate a case which is characterized by its fixed energy $E^* = 2$. For this case equation (2.33) takes the form of

$$\dot{\phi} = \pm \omega_0 \sqrt{2(\cos(\phi) + 1)}. \tag{2.34}$$

Substituting $\cos(\phi) = y$, we get

$$\omega_0 \int_0^t dt = \int_y^1 \frac{dy'}{\sqrt{(1 - y'^2)(1 + y')}}. \tag{2.35}$$

The integration of this equation yields

$$\pm \omega_0 \sqrt{2} t = -\sqrt{2} \text{Arctanh} \left\{ \frac{\sqrt{(1 + y)(1 - y^2)}}{\sqrt{2}(1 + y)} \right\}. \tag{2.36}$$

By inverting these functions, the solution for the angle ϕ is obtained

$$\phi = \arccos \left(1 - 2 \tanh^2(\omega_0 t) \right). \tag{2.37}$$

From (2.33), we get for $0 \le E^* < 2$

$$\int_0^t dt' = \pm \frac{1}{\sqrt{2}\omega_0} \int_0^\phi \frac{d\phi'}{\sqrt{\cos(\phi') - (1 - E^*)}}. \tag{2.38}$$

If we replace $1 - E^* = \cos(\phi_1)$ and $k^2 = \sin(\phi_1/2)$, we can express (2.38) in the form of

$$\pm \omega_0 t = \int_0^y \frac{dy'}{\sqrt{(1 - y'^2)(1 - k^2 y'^2)}} = \mathrm{sn}^{-1}(y, k). \qquad (2.39)$$

where sn is the inverse function of the Jacobian elliptic function, in *Mathematica* known as **JacobiSN[ ]**, and leads to

$$y = \mathrm{sn}(\omega_0 t, k). \qquad (2.40)$$

With respect to angle ϕ, we get the expression

$$\phi = 2 \arcsin \left(k \, \mathrm{sn}(\omega_0 t, k) \right). \qquad (2.41)$$

If we choose $E^* > 2$, we obtain the solution for the angle by applying a similar strategy to the one above. The solution is

$$\phi = 2 \, \mathrm{am}(\omega_0 t, k) \qquad (2.42)$$

where **am** denotes the **JacobiAmplitude[ ]**. The course of the solutions for the various k values is $k = \{0.1, 0.5, 0.9\}$; i.e., different initial amplitudes and $\omega_0 = 4$ are shown in figures 2.6, 2.7 and 2.8. The figures are produced with **Plot[ ]** as well as with **ArcSin[ ]**, **JacobiSN[ ]** and **JacobiAmplitude[ ]**. The Jacobi elliptic functions have two arguments, the independent variable $\omega_0 t$ and the modulus k.

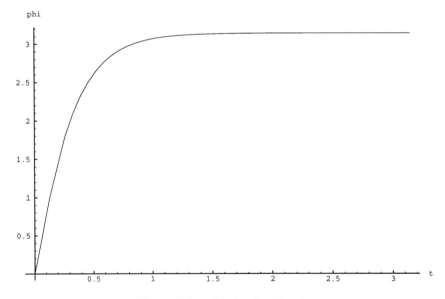

Figure 2.6.   Solution for $E^* = 2$.

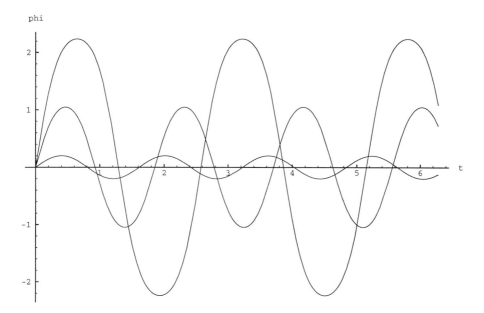

Figure 2.7. Solutions for $0 \leq E^* < 2$. The amplitudes of the solution increase by increasing the values of the modulus $k = \{0.1, 0.5, 0.9\}$.

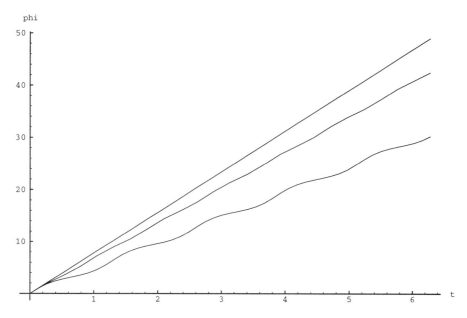

Figure 2.8. Solutions of the mathematical pendulum for $E^* > 2$. The slope of the solution decreases by increasing the modulus k. The three values for k are $\{0.1, 0.5, 0.9\}$.

2.4 The damped driven pendulum

Another familiar example is the planar pendulum subject to a driving force and frictional damping. The motion of the damped, driven pendulum is described by the equation

$$\ddot{y} + \alpha\dot{y} + \frac{g}{l}\sin(y) = \gamma\cos(\omega t) \tag{2.43}$$

Apart from its application to the pendulum, this equation describes a Josephson tunneling junction in which two superconducting materials are separated by a thin nonconducting oxide layer. Among the practical applications of such junctions are high–precision magnetometers and standards of voltage elements. The ability of these Josephson junctions to switch rapidly and with very low dissipation from one current-carrying state to another may provide microcircuit technologies for, say, supercomputers, that are more efficient than those based on conventional semiconductors. Hence the nature of the dynamic response of a Josephson junction to the external driving force—the $\cos(\omega t)$ term—is a matter of technological as well as of fundamental interest.

One of the characteristics of this equation is the occurrence of chaotic states. These states depend on the choice of parameters for damping and driving force. Since standard analytical techniques are of limited use in the chaotic regime, we demonstrate the existence of chaos by relying on graphical results from numerical simulations.

We first note that since there is an external time dependence in the equation of motion, the system really involves three first order differential equations. In a normal dynamic system, each degree of freedom results in two first–order equations and such a system is said to correspond to one–and–a–half degrees of freedom. To see this explicitly, we introduce the variable $\mathbf{z} = \boldsymbol{\mathrm{omega}}\ \mathbf{t}$ and rewrite the equation of motion resulting in

```
D[y[t],t]  == p/(m l^ 2)

D[p[t],t]  == -alpha p - mgl Sin[y] + m l^ 2 gamma Cos[z]

D[z[t],t]  == omega
```

The equations show how the system depends on the three generalized coordinates: y, p, and z. Note further that the presence of damping implies that the system is no longer Hamiltonian but is dissipative and thus can have attractors.

Analysis of the damped driven pendulum illustrates two separate but related aspects of chaos: first, the existence of a strange attractor and second, the presence of several different attracting sets and the resulting extreme sensitivity of the asymptotic motion to initial conditions.

To identify the signature of chaos we use the Poincaré technique to represent a section of phase space. A Poincaré section is a plot showing only the phase-plane variables **y** and **D[y,t]**. A stroboscopic snapshot of the motion is taken during each cycle of the driving force. The obtained complicated attracting set of points shown in figure 2.14 is in fact a strange attractor and describes a never–repeating, non–periodic motion in which the pendulum oscillates and flips over its pivot point in an irregular, chaotic manner. Before we examine this chaotic behavior let us first discuss the regular motion of the system.

2.4.1 Regular motion

We use for the numerical integration a slightly different system of equations than given above. The relevant system of equations reads

```
f1[y_,p_] := p;

f2[y_,p_] := -w0^2 Sin[y] -a p + b Cos[w t];
```

We set the constants and parameters of these equations to

```
Needs["ProgrammingExamples`RungeKutta`"];

w0 = 1; a = 0.2; b= 0.52; w=0.694;

cycl = 30; steps = 31;
```

To integrate the equations of motion, the fourth order Runge Kutta procedure with finite step-size is used

```
pts = RungeKutta[ {f1[y,p], f2[y,p]}, {y,p},{0.8,0.8},
                {t,0,cycl(2Pi/w),(2Pi/w)/steps} ];
```

The result of the integration procedure (see figure 2.9) is now displayed

```
Show[ Graphics[{ Line[pts],
            Text[y,{6.5,0.2}], Text[p,{1.2,2}] }],
        Axes->True, PlotRange->{-2.2,2.},
        PlotLabel->"driven pendulum" ];
```

driven pendulum

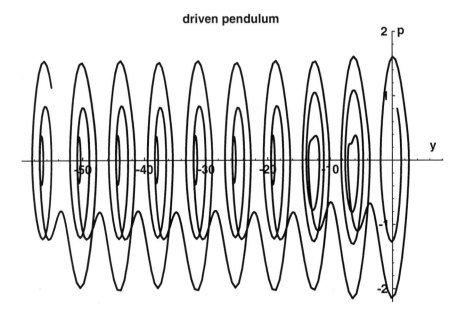

Figure 2.9.   Phase space representation of a trajectory for the driven pendulum.

The mathematical solution of the equations is obtained in the range for y of **[-Infinity,Infinity]**. However, since the real motion of a pendulum is restricted to the range **[-Pi,Pi]**, we need to reduce the total time period to this interval. To find the motion modulo **2Pi**, we define the function

```
pi:=Pi//N;

red[x_] :=  Mod[x,2pi]      /; Mod[x,2pi] <= pi;

red[x_] := (Mod[x,2pi]-2pi) /; Mod[x,2pi] >  pi;
```

and map this function onto the first argument of each of the solutions **pts**

```
xpos = Table[{i,1}, {i,Length[pts]} ];

npts = MapAt[ red, pts, xpos];

mpts = Take[npts, {20 steps,cycl steps} ];
```

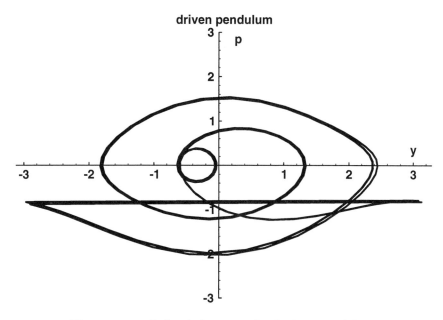

Figure 2.10.   Reduced phase space for the driven pendulum.

From **npts** we extract only those points which are in a steady oscillation state
(figure 2.10)

```
Show[ Graphics[ {Line[mpts],
                Text[y,{3,0.3}], Text[p,{.3, 2.8}] }],
      Axes->True, PlotRange->{-3.,3.},
      PlotLabel->"driven pendulum" ];
```

To show the oscillating behavior of this solution a Poincaré section is created
by a stroboscopic map (figure 2.11). We extract only those points of the solution
which are commensurate with the driving frequency

```
poincare = Table[ npts[[n]],
                {n,1+10 steps, Length[npts],steps} ];

Length[poincare]

21

ListPlot[ poincare, PlotRange->{{-.5,2},{-1.5,1.5}},
          PlotStyle->{RGBColor[1,0,0],PointSize[0.011]},
          AxesLabel->{" y"," p"} ,
          PlotLabel->"          Poincare section" ];
```

2.4.2 Chaotic behavior

As a second example we consider the damped driven pendulum with parameters $a = 1/2$, $b = 1.15$ and $w = 2/3$. Initial conditions are again $y = 0.8$ and $p = 0.8$. The procedure to reduce this data is the same as in the example above, the result is shown in figure 2.12.

```
f1[y_,p_] := p;

f2[y_,p_] := -w0^2 Sin[y] -a p + b Cos[w t];

w0 = 1; a = 1/2; b= 1.15; w=2/3;

cycl = 500; steps = 30;

ptsc = RungeKutta[ {f1[y,p], f2[y,p]}, {y,p},{0.8,0.8},
                    {t,0,cycl(2Pi/w),(2Pi/w)/steps} ];

Show[ Graphics[{ Line[ptsc],
                Text[y,{20,0.3}], Text[p,{3.,2.8}] }],
        Axes->True, PlotRange->{-3,3},
        PlotLabel->"driven pendulum chaotic state" ];
```

A reduction of the phase space to the interval **[-Pi,Pi]** gives us the impression of a chaotic state

```
xposc = Table[{i,1}, {i,Length[ptsc]} ];

nptsc = MapAt[ red, ptsc, xposc];

mptsc = Take[nptsc, {20 steps,cycl steps} ];

Show[ Graphics[ {Line[mptsc],
                Text[y,{3,0.3}], Text[p,{.3, 2.8}] }],
        Axes->True, PlotRange->{-3.,3.},
        PlotLabel->"reduced phase space" ];
```

2.4.2.1 Poincaré section A convenient way to delineate the dynamics of a system is given by the Poincaré section. The Poincaré section represents a slice of the phase space of the system. For the three-dimensional case under examination, a slice can be obtained from the intersection of a continuous trajectory with a two-dimensional plane in the phase space. One method of creating a Poincaré section is to check the system over a full cycle of the driving frequency. If we are dealing with a periodic evolution of period n, then this sequence consists of n dots being indefinitely repeated in the same order (compare figure 2.11). If the evolution is

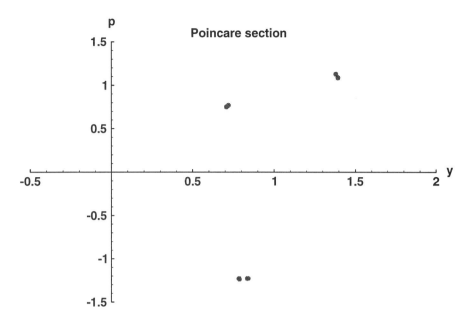

Figure 2.11.   Poincaré section of the driven pendulum for a periodic solution.

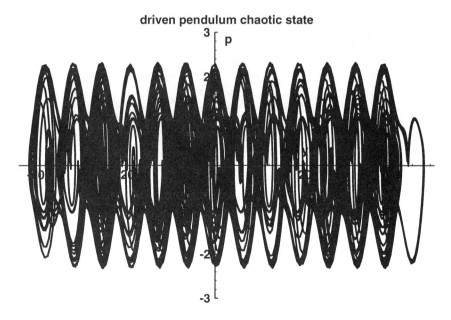

Figure 2.12.   Phase space representation of the driven pendulum in a chaotic state.

chaotic, then the Poincaré section is a collection of points that show interesting patterns with no obvious repetition (compare figure 2.14). The process of obtaining a Poincaré section can be compared to sampling the state of the system randomly instead of continuously. An example for the damped driven pendulum is given below.

```
poincarec = Table[ nptsc[[n]],
                    {n,1+20 steps, Length[nptsc],steps} ];

Length[poincarec]

481

ListPlot[ poincarec, PlotRange->{{-4.,4},{-3.,3.}},
          PlotStyle->{RGBColor[1,0,0],
                      AbsolutePointSize[1.85]},
          AxesLabel->{" y"," p"} ,
          PlotLabel->"Poincare section" ];
```

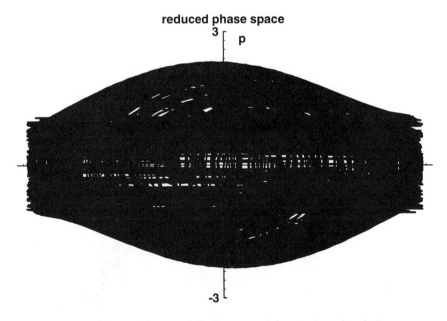

Figure 2.13.   Chaotic behavior of the driven pendulum in the reduced phase space.

We observe from figure 2.13 that the motion in phase space takes place on a finite attracting subset. This subset of phase space has a characteristic shape depending on the parameters used in the integration process. The complicated attracting set shown in figure 2.14 is in fact a strange attractor and describes a

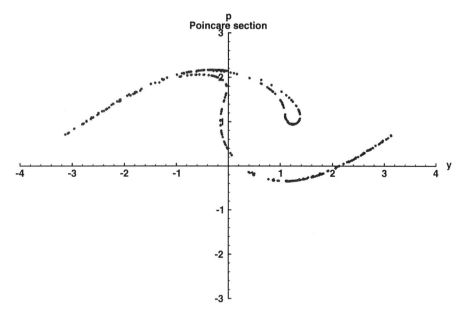

Figure 2.14. Strange attractor of the driven pendulum.

never repeating, non-periodic motion in which the pendulum oscillates and flips over its pivot point in an irregular, chaotic manner.

2.5 Linear chain

In contrast to our earlier example in section 2.1, the linear chain is a more complicated system which consists of $N+1$ springs connecting identical particles with each other. As before, we assume that the end points of the springs are fixed. Let the mass of each particle be m and the equilibrium distance between them d. Thus, the total length of the chain is $L = (N+1)d$. We will study the case of small transverse oscillations. For this purpose we are investigating the neighboring particles $j-1$, j and $j+1$ of which j can be any particle within the chain. We assume that x_{j-1}, x_j, and x_{j+1} are small. The spring constant κ is the same for each spring.

The force restoring a particle into its equilibrium is given by

$$F_j = -\kappa(x_j - x_{j-1}) - \kappa(x_j - x_{j+1}). \tag{2.44}$$

In accordance with Newton's Law, the equation of motion for the j-th mass is

$$\ddot{x}_j = \frac{\kappa}{m}(x_{j-1} - 2x_j + x_{j+1}). \tag{2.45}$$

where the boundary conditions are $x_0 = x_{N+1} = 0$. Thus, the equations of motion (2.45) follow from $\dot{p}_j = \{p_j, H\}$ and $\dot{x}_j = \{x_j, H\}$. If we eliminate the momenta $p_j = m\dot{x}_j$ in this set of equations, we obtain (2.45). Note that each particle is connected to only one nearest neighbor. This relationship is known as nearest neighbor reciprocity.

The equations of motion equivalent to (2.45) can be derived with the help of the Poisson bracket. The total energy or the Hamiltonian of the chain is

$$H = \sum_{j=1}^{N} \frac{p_j^2}{2m} + \frac{\kappa}{2} \sum_{j=1}^{N+1} (x_{j-1} - x_j)^2 \qquad (2.46)$$

which is the sum of kinetic and potential energy of the particles.

In order to find a solution to (2.45), we substitute the expression

$$x_j(t) = a_j e^{i\omega t} \qquad (2.47)$$

into (2.45) and get

$$-\kappa a_{j-1} + \left(2\kappa - m\omega^2\right) a_j - \kappa a_{j+1} = 0 \qquad \text{with} \qquad j = 1, 2, \ldots, n. \qquad (2.48)$$

The amplitudes a_j are complex valued quantities. Since the end points of the chain are firmly fixed, we also need to satisfy $a_0 = a_{N+1} = 0$. Equation (2.48) represents a homogeneous linear system of equations for amplitudes $(a_1, a_2, \ldots, a_N)$. For non–disappearing amplitudes, the determinant of the coefficient matrix must be equal to zero in order to yield N eigenfrequencies ω^2.

The secular equation in the form of a polynomial in ω^2 of N-th degree is easy to solve for a small number of particles N whereas for large N the solution becomes complicated. If we have a large number of chain members, it is easier to use the following method. As a solution for the amplitudes, we use the harmonic term

$$a_j = a e^{i(j\gamma + \delta)} \qquad (2.49)$$

with a real amplitude a. Inserting this expression into (2.48), we get

$$-\kappa e^{-i\gamma} + \left(2\kappa - m\omega^2\right) - \kappa e^{i\gamma} = 0. \qquad (2.50)$$

Solving for ω^2, we find

$$\omega^2 = \frac{2\kappa}{m} - \frac{\kappa}{m}\left(e^{i\gamma} + e^{-i\gamma}\right) \qquad (2.51)$$

$$= \frac{2\kappa}{m}(1 - \cos(\gamma)) \qquad (2.52)$$

$$= \frac{4\kappa}{m}\sin^2\left(\frac{\gamma}{2}\right). \qquad (2.53)$$

Since the secular determinant is of N-th order, N solutions must exist for ω. Hence, we can write

$$\omega_\nu = 2\sqrt{\frac{\kappa}{m}} \sin\left(\frac{\gamma_\nu}{2}\right) \qquad \nu = 1, 2, \ldots, N. \tag{2.54}$$

We still need to determine the parameters γ_ν and δ_ν, which can be obtained from the boundary conditions. We have

$$a_{j\nu} = a_\nu \cos(j\gamma_\nu - \delta_\nu) \tag{2.55}$$

on the condition that $a_{0,\nu} = a_{(N+1)\nu} = 0$. In order for (2.55) to satisfy the end points, $\delta_\nu = \pi/2$, that is

$$\begin{aligned}
a_{j\nu} &= a_\nu \cos\left(j\gamma_\nu - \frac{\pi}{2}\right) \\
&= a_\nu \sin(j\gamma_\nu).
\end{aligned} \tag{2.56}$$

The following equation applies to the other end point

$$a_{(N+1)\nu} = 0 = a_\nu \sin((N+1)\gamma_\nu) \tag{2.57}$$

or

$$(N+1)\gamma_\nu = \nu\pi \qquad \nu = 1, 2, \ldots, N \tag{2.58}$$

and hence

$$\gamma_\nu = \frac{\nu\pi}{N+1}. \tag{2.59}$$

Since there are only N different ω_ν's available as solutions, we get

$$a_{j\nu} = a_\nu \sin\left(j\frac{\nu\pi}{N+1}\right) \tag{2.60}$$

for the amplitudes.

Thus, the general solution takes the form

$$\begin{aligned}
x_j &= \sum_{\nu=1}^{N} \tilde{c}_\nu a_{j\nu} e^{i\omega_\nu t} \\
&= \sum_{\nu=1}^{N} \beta_\nu \sin\left(j\frac{\nu\pi}{N+1}\right) e^{i\omega_\nu t},
\end{aligned} \tag{2.61}$$

where the frequencies ω_ν are given by

$$\omega_\nu = 2\sqrt{\frac{\kappa}{m}} \sin\left(\frac{\nu\pi}{2(N+1)}\right). \tag{2.62}$$

Relevant to our purposes is the real part of the solution x_j which, using $\beta_j = b_j + ic_j$, is given via

$$x_j(t) = \sum_{\nu=1}^{N} \sin\left(j\frac{\nu\pi}{N+1}\right)(b_\nu \cos(\omega_\nu t) - c_\nu \sin(\omega_\nu t)). \tag{2.63}$$

These general relationships can be reduced to a case of two particles if $N = 2$ and κ is replaced with $k = k_{12}$. To illustrate the time history of the oscillations, we first need to determine the complex expansion coefficients $\beta_{j\nu}$ in (2.63). This is done by specifying the initial conditions $x_j(t = 0) = x_j^0$ and $\dot{x}_j(t = 0) = \dot{x}_j^0$. From these conditions, we obtain a system of coupled linear equations of the form

$$x_j^0 = \sum_{\nu=1}^{N} b_\nu \sin\left(j\frac{\nu\pi}{N+1}\right) \tag{2.64}$$

$$\dot{x}_j^0 = -\sum_{\nu=1}^{N} \omega_\nu c_\nu \sin\left(j\frac{\nu\pi}{N+1}\right) \qquad j = 1, 2, \ldots, N. \tag{2.65}$$

Since β_ν is a complex quantity with $\beta_\nu = b_\nu + ic_\nu$, we obtain a system of equations with $2N$ unknowns.

To show how to construct a package, we summarize our theoretical results in a few *Mathematica* functions. In our first example, we limit our interest to the presentation of oscillations.

Since the frequencies ω_ν are determined by (2.62) and enter our solutions as characteristic parameters, we start with defining the frequency. First, the initial conditions x_j^0 and $\dot{x}_j^0$ can be combined in one list **InitialConditions**. The set–up of this list is as follows: the first N elements contain x_j^0's where $\dot{x}_j^0$ begins with the $N + 1$st element. With this list we can solve (2.64) and (2.65) by inserting the results b_ν and c_ν into relation (2.63).

In addition to the constants b_ν and c_ν, the frequency also functions as a parameter in (2.63). The frequency itself depends on two characteristic parameters, index ν and the total number of particles N. For simplicity sake we set $\kappa = \tau/d$ and assume $\tau = d = m = 1$ and define **omega** as follows:

```
omega[nue_,n_] := Block[{tau=1, d=1, m=1},
            2 Sqrt[tau/(m d)] Sin[(nue Pi)/(2 (n+1))]
            ]
```

(For clarity, we are suppressing input and output labels at this point.) In our next step, we derive the right hand sides of equations (2.64) and (2.65) both of which contain b_ν and c_ν as unknown variables. Index j and the number of particles N are necessary for the input of

```
Initx[j_,n_] :=Sum[b[nue] Sin[j nue Pi/(n+1)],
                  {nue,1,n}]

Initxdot[j_,n_] := -Sum[omega[nue,n] c[nue]
                           Sin[j nue Pi/(n+1)],
                    {nue,1,n}]
```

With **Initxdot**[] we have used **omega** as a function. The first argument of **Sum**[] contains the addend. The second argument contains the range of summation which is included in braced brackets. Equations (2.64) and (2.65) can now be set up in the form of

```
eqns = Thread[InitialConditions ==
          Transpose[
              Table[{Initx[j,n],Initxdot[j,n]},
              {j,1,n}]]]
```

The **Equal** function denoted by == signifies the comparison of two expressions. True is returned if the equation is satisfied. Before finding solutions for the unknown variables with the help of **Solve**[], we first have to create the unknown variables b_ν and c_ν. The list of variables is set up as follows:

```
vars = Table[{b[nu],c[nu]},{nu,1,n}] // Flatten
```

Table[] produces a list whose elements are defined by its first argument. The second argument, separated by a comma, indicates the range in which the variable **nu** is found. The structure of this function is identical to the structure of the functions and **Sum**[]. We eliminate the extra brackets of the list **vars** by applying **Flatten**[] to the nested list. Consequently, we get the solutions of equations (2.64) and (2.65) with

```
Result = Solve[eqns,vars]
```

Since the result is obtained in the form of a replacement rule, we are able to use it immediately in our equation (2.63). In order to produce the solutions to (2.63), we define the following

```
x[j_,n_] := Sum[(b[j,nue] + I c[j,nue]) Sin[j nue Pi/(n+1)]
                Exp[I omega[nue,n] t],
             {nue,1,n}]
```

As a result, all N solutions are

```
Solution = Table[x[j,n],{j,1,n}]
```

We use the results from the initial conditions as follows

```
Solution = Solution /. Result
```

to obtain a list which contains all x_j. Finally, we need to specify the time t for evaluating the motion. This can be done by

```
List1 = Solution /. t->0
```

For $t = 0$, a presentation of the solution of the time factor $t = 0$ is given by

```
ListPlot[List1]
```

Hence, the time history of the oscillation has been created. By varying the time, we can create the time history of the movement of the particles as follows

```
Do[
   ListPlot[Solution /. t->t0, {t0,0,2,0.1}]
   ]
```

Twenty consecutive presentations result which differ in time by $\Delta t = 0.1$. The functions defined above are joined in the package **LinearChain'** listed below.

```
                          ─── File linearc.m ───
 BeginPackage["LinearChain`"]

 Clear[omega, Anfx, Anfxdot, x, ChainPlot]
 (* --- Clear removes the variables and functions given in the argument --- *)

 ChainPlot::usage = "ChainPlot[Initial_List,tend_] creates a plot of the chain
 for 10 time steps. Input variables are a list of coordinates and velocities
 for the initial time t0. The second argument is the end point of the
 time. As output we get 10 pictures stored in the variables pl[1]-pl[10]."

 Begin["`Private`"]

 (* --- frequency of the chain --- *)

  omega[nu_,n_,tau_:1,d_:1,m_:1] :=2*Sqrt[tau/(m*d)]*Sin[(nu*Pi)/(2*(n+1))]

 (*--- right hand side of the equation for x related to
       the initial conditions ---*)

  Inix[j_,n_] := Sum[b[nu]*Sin[j*nu*Pi/(n+1)],
                  {nu,1,n}]

 (*--- right hand side of the equation for x' related to
       the initial conditions ---*)

  Inixdot[j_,n_] := -Sum[omega[nu,n]*c[nu]*
                        Sin[j*nu*Pi/(n+1)],
                     {nu,1,n}]


 (* --- representation of the real part of x --- *)

  x[j_,n_] := Block[{nu},
               Sum[Sin[j*nu*Pi/(n+1)]*(b[nu]*Cos[omega[nu,n]*t] -
                  c[nu]*Sin[omega[nu,n]*t]),
               {nu,1,n}]
               ]

 (* --- Plot of the chain movement --- *)

  ChainPlot[Initial_List,tend_]:=Block[
                           {n, k=0, gl1={}, gl2={}, var={}, Result,
                            Solution,dt},
       n = Length[Initial]/2;
 (* --- create equations --- *)
 eqns = Thread[InitialConditions ==
            Transpose[
                Table[{Initx[j,n],Initxdot[j,n]},
                {j,1,n}]]];

 (* --- create a list of variables --- *)
 vars = Table[{b[nu],c[nu]},{nu,1,n}] // Flatten;

 (* --- solve equations ---- *)
       Result = Solve[eqnsl,vars];
```

```
      Result = Flatten[Result];
(* --- create the general solution --- *)
      Solution = Table[x[j,n],{j,1,n}];
(* --- replace the results in the general solution --- *)
      Solution = Solution /. Result;
(* --- output the results in plots ---- *)
      dt = tend*0.1;
      Do[
        k = k + 1;
        pl[k] = ListPlot[Solution /. t->t0,
                    PlotJoined->True,
                    AxesLabel->{"n","x"}],
        {t0,0,tend,dt}];
      plot = Table[pl[i],{i,1,10}];
      Show[plot]
      ]
  End[]
  EndPackage[]
```

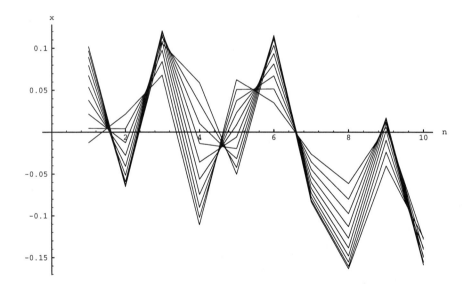

Figure 2.15.   Time history of the oscillation of a chain consisting of ten particles at random initial conditions.

Figure 2.15 shows the time history of the oscillation of a chain consisting of ten particles. The initial conditions are given by

```
InitialCondition = Table[Random[Real,{-.2,.2}],{i,1,20}]
```

Function **Random**[ ] yields real random numbers which are equally distributed between -0.2 and 0.2. The drawing is made with **ChainPlot[InitialCondition,1]**.

2.6 The two body problem

So far we have examined a couple of problems describing the motion of few and large numbers of particles. In this section we return to a system consisting of two particles.

We discuss the motion of a system consisting of two bodies which interact via central forces. The problem of the interaction of two bodies is of central importance to physics because it allows us to describe the motion and interaction of two large bodies like the moon and the earth. The model further allows us to discuss the classical motion of two atoms or the electron in an atom.

From a mathematical point of view, the two body problem is one of the rare problems which has an analytical solution. However, in our examination we do not solve the problem by analytical methods but use numerical techniques to derive a representation of the solution. Our main goal is to find the solutions which describe the motion in Cartesian coordinates. By deriving the solutions, we also illustrate the animation capabilities of *Mathematica* in the simulation of real space motion.

One of the essential steps in the solution process of the two body problem is the introduction of an appropriate system of coordinates. To describe the motion in an efficient and simple form, we introduce center of mass and relative coordinates. To follow the real path of the two bodies in time, we need six coordinates which are given by the three Cartesian coordinates of each body. The coordinates themselves depend on time. We first introduce the two vectors r_1 and r_2 which specify the location of the bodies in space (see figure 2.16).

```
<<Graphics`Arrow`

vectors={Arrow[{0,0},{1,2}],Arrow[{0,0},{3,-2}],
        Arrow[{0,0},{2,0}],Arrow[{3,-2},{1,2}]};
text={Text[FontForm["center of mass",{"Helvetica",8}],
                {2.5,0}],
    Text[FontForm["m1",{"Helvetica",8}],{1.25,2}],
    Text[FontForm["m2",{"Helvetica",8}],{3.25,-2}],
    Text[FontForm["r1",{"Helvetica",8}],{0.5,0.75}],
    Text[FontForm["r2",{"Helvetica",8}],{0.5,-0.5}],
    Text[FontForm["R",{"Helvetica",8}],{1.5,0.25}],
    Text[FontForm["r",{"Helvetica",8}],{2.5,-0.5}]};
Show[Graphics[{vectors,text}]]


-Graphics-
```

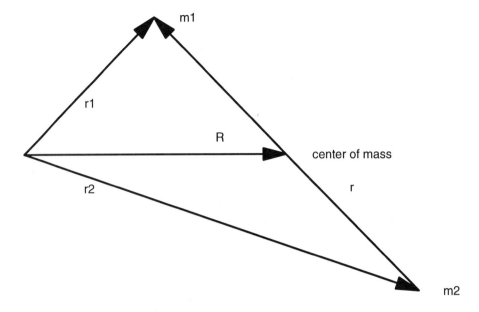

Figure 2.16.   Geometrical location of the two bodies.

```
r1 = {x1[t],y1[t],z1[t]};

r2 = {x2[t],y2[t],z2[t]};
```

It is useful to specify the positions of the two particles of masses m_1 and m_2 with respect to the center of mass. We define the center of mass vector $\boldsymbol{R}$ as

```
R = (m1 r1 + m2 r2)/(m1 + m2)

  m1 x1[t] + m2 x2[t]   m1 y1[t] + m2 y2[t]
{--------------------, --------------------,
       m1 + m2               m1 + m2
  m1 z1[t] + m2 z2[t]
  --------------------}
       m1 + m2
```

The second vector simplifying the description of motion is given in relative coordinates r of the two particles as

```
r = r1 - r2

{x1[t] - x2[t], y1[t] - y2[t], z1[t] - z2[t]}
```

Each of these relations establishes a transformation from the Cartesian coordinates r_1 and r_2 to a particle related coordinate system. The two representations of the coordinates are equivalent and invertible. Using the transformations in the form given above, we are able to simplify the representation of the equations of motion and find their solutions in real space.

2.6.1 Equations of motion

According to Newton's law, if particle one acts by a force on particle two, then particle two acts by the inverse force on particle one. This mutual interaction of forces governs the equations of motion. We consider a special type of force depending only on the inter–particle distance called a central force. The behavior of the central force leads to the equations of motion by applying Newton's second law. The motion of the first particle is described by

$$\ddot{r}_1 + m_2 \frac{r_1 - r_2}{|r_1 - r_2|^2} = 0 \quad \text{and} \tag{2.66}$$

$$\ddot{r}_2 + m_1 \frac{r_2 - r_1}{|r_1 - r_2|^2} = 0. \tag{2.67}$$

These two relations are represented in *Mathematica* by

```
gl1 = D[r1,{t,2}]  +
     m2 (r1-r2)/((r1-r2).(r1-r2))^(3/2)

{(m2 (x1[t] - x2[t])) /
                   2                2
     Power[(x1[t] - x2[t])  + (y1[t] - y2[t])  +
                   2
       (z1[t] - z2[t]) , 3/2] + x1''[t],
   (m2 (y1[t] - y2[t])) /
                   2                2
     Power[(x1[t] - x2[t])  + (y1[t] - y2[t])  +
                   2
       (z1[t] - z2[t]) , 3/2] + y1''[t],
   (m2 (z1[t] - z2[t])) /
                   2                2
     Power[(x1[t] - x2[t])  + (y1[t] - y2[t])  +
                   2
       (z1[t] - z2[t]) , 3/2] + z1''[t]}
```

The equation of motion for the second particle reads

```
gl2 = D[r2,{t,2}] +
      m1 (r2-r1)/((r2-r1).(r2-r1))^(3/2)

{(m1 (-x1[t] + x2[t])) /
                         2                    2
     Power[(-x1[t] + x2[t])  + (-y1[t] + y2[t])  +
                        2
     (-z1[t] + z2[t]) , 3/2] + x2''[t],
  (m1 (-y1[t] + y2[t])) /
                         2                    2
     Power[(-x1[t] + x2[t])  + (-y1[t] + y2[t])  +
                        2
     (-z1[t] + z2[t]) , 3/2] + y2''[t],
  (m1 (-z1[t] + z2[t])) /
                         2                    2
     Power[(-x1[t] + x2[t])  + (-y1[t] + y2[t])  +
                        2
     (-z1[t] + z2[t]) , 3/2] + z2''[t]}
```

As mentioned above, it is essential to express the locations of the bodies r_1 and r_2 by means of center of mass coordinates and relative coordinates. The real space coordinates r_1 and r_2 are expressible in these coordinates by applying the rules

```
s3 = Thread[
       {x1,y1,z1}->{Function[t,R1[t] + m2 x[t]/(m1 + m2)],
                    Function[t,R2[t] + m2 y[t]/(m1 + m2)],
                    Function[t,R3[t] + m2 z[t]/(m1 + m2)]}
       ]

                          m2 x[t]
{x1 -> Function[t, R1[t] + -------],
                          m1 + m2
                          m2 y[t]
   y1 -> Function[t, R2[t] + -------],
                          m1 + m2
                          m2 z[t]
   z1 -> Function[t, R3[t] + -------]}
                          m1 + m2
```

Rule s3 inverts the relation for R and r. To denote the center of mass coordinates, we use R1, R2 and R3 as coordinate components. The components of the relative vector are denoted by x, y and z. A similar representation of the transformation from Cartesian to particle related coordinates is possible for the second mass

```
s4 = Thread[
    {x2,y2,z2}->{Function[t,R1[t] - m1 x[t]/(m1 + m2)],
                 Function[t,R2[t] - m1 y[t]/(m1 + m2)],
                 Function[t,R3[t] - m1 z[t]/(m1 + m2)]}
    ]

                          m1 x[t]
{x2 -> Function[t, R1[t] - -------],
                          m1 + m2
                          m1 y[t]
 y2 -> Function[t, R2[t] - -------],
                          m1 + m2
                          m1 z[t]
 z2 -> Function[t, R3[t] - -------]}
                          m1 + m2
```

The center of mass coordinates are denoted again by **R1**, **R2** and **R3** and the relative coordinates by **x**, **y** and **z**. Applying the transformations **s3** and **s4** to the equations of motion stored in **gl1** and **gl2**, we obtain a representation of the equations in center of mass coordinates and relative coordinates. The equation of the first body **gl1** is then given by

```
gl3 = gl1 /. s3 /. s4 //Simplify

          m2 x[t]                      m2 x''[t]
{------------------------- + R1''[t] + ---------,
      2     2     2 3/2                 m1 + m2
 (x[t]  + y[t]  + z[t] )
          m2 y[t]                      m2 y''[t]
------------------------- + R2''[t] + ---------,
      2     2     2 3/2                 m1 + m2
 (x[t]  + y[t]  + z[t] )
          m2 z[t]                      m2 z''[t]
------------------------- + R3''[t] + ---------}
      2     2     2 3/2                 m1 + m2
 (x[t]  + y[t]  + z[t] )
```

The second equation of motion **gl2** is transformed to

```
gl4 = gl2 /. s3 /.s4 //Simplify

          m1 x[t]
{-(-------------------------) + R1''[t] -
      2     2     2 3/2
 (x[t]  + y[t]  + z[t] )
 m1 x''[t]            m1 y[t]
 ---------, -(-------------------------) +
```

```
    m1 + m2              2       2       2 3/2
                    (x[t]  + y[t]  + z[t] )
              m1 y''[t]
    R2''[t] - ---------,
              m1 + m2
              m1 z[t]
  -(------------------------) + R3''[t] -
         2       2       2 3/2
     (x[t]  + y[t]  + z[t] )
     m1 z''[t]
     ---------}
     m1 + m2
```

Both equations contain similar terms with opposite signs. In the present form the two equations are obviously coupled. To decouple both types of equations, we have to add or subtract the transformed equations **gl3** and **gl4**. Subtraction of **gl3** from **gl4** and replacing mass **m1** by the total mass **M** delivers the equation of motion for the relative coordinates

```
  glr = (gl3-gl4)/.m1->M-m2//Simplify

            M x[t]
  {-------------------------- + x''[t],
        2       2       2 3/2
    (x[t]  + y[t]  + z[t] )
            M y[t]
  -------------------------- + y''[t],
        2       2       2 3/2
    (x[t]  + y[t]  + z[t] )
            M z[t]
  -------------------------- + z''[t]}
        2       2       2 3/2
    (x[t]  + y[t]  + z[t] )
```

The equations of motion describing the pure center of mass motion are derivable by multiplying the first equation **gl3** by **m2** and the second **gl4** by **m1**. Adding these two equations together and replacing mass **m1** by the total mass **M** yields the center of mass equations

```
  gls = (gl3 m1 + m2 gl4) /. m1->M-m2//Simplify

  {M R1''[t], M R2''[t], M R3''[t]}
```

The solution of the center of mass equations is straightforward since we have linear second order equations of the form

```
gls1 = Thread[gls=={0,0,0}]

{M R1''[t] == 0, M R2''[t] == 0, M R3''[t] == 0}
```

The solution follows from applying the function **DSolve[ ]**

```
sgls = DSolve[gls1,{R1,R2,R3},t]

{{R1 -> Function[t, C[1] + t C[4]],
  R2 -> Function[t, C[2] + t C[5]],
  R3 -> Function[t, C[3] + t C[6]]}}
```

The solutions show that the motion of the center of mass is free of acceleration and follows a straight line in space. This means that the center of mass is translated linearly with time.

So far, we have derived the solution of the center of mass motion. We examine next the motion of particles related to the relative coordinates.

Since angular momentum is conserved and the planetary motion is bound to a plane, we are free to introduce plane polar coordinates to describe the motion in angular and radial components. The transformation from relative coordinates to radial and angular components is given by rule **s6**.

```
Clear[r]

s6 = Thread[
     {x,y,z}->{Function[t,r[t] Cos[Phi[t]] ],
               Function[t,r[t] Sin[Phi[t]] ],
               Function[t,0]}
     ]

{x -> Function[t, r[t] Cos[Phi[t]]],
  y -> Function[t, r[t] Sin[Phi[t]]],
  z -> Function[t, 0]}
```

Applying rule **s6** to the equations **glr** we find the equations of the radial and angular components.

```
glr1 = glr /. s6//PowerExpand//Simplify

  M Cos[Phi[t]] r[t]
{------------------ - 2 Sin[Phi[t]] Phi'[t] r'[t] -
        2 3/2
```

```
   (r[t] )
                        2
  r[t] (Cos[Phi[t]] Phi'[t] +
      Sin[Phi[t]] Phi''[t]) + Cos[Phi[t]] r''[t],
 M r[t] Sin[Phi[t]]
 ------------------ +
          2 3/2
     (r[t] )
  2 Cos[Phi[t]] Phi'[t] r'[t] +
                               2
  r[t] (-(Sin[Phi[t]] Phi'[t] ) +
      Cos[Phi[t]] Phi''[t]) + Sin[Phi[t]] r''[t], 0}
```

The resulting list of equations contains coupled differential equations of second order describing the motion in **Phi** and **r**. If we compare the coefficients of the linearly independent trigonometric functions **Cos[]** and **Sin[]** in **glr1**, we observe that despite their signs the two equations are equivalent. We extract the two independent equations by

```
 glr11 = Coefficient[Expand[glr1],Cos[Phi[t]]]

              2
  M Sqrt[r[t] ]                2
 {------------- - r[t] Phi'[t]  + r''[t],
          3
      r[t]
   2 Phi'[t] r'[t] + r[t] Phi''[t], 0}

 glr12 = Coefficient[Expand[glr1],Sin[Phi[t]]]

 {-2 Phi'[t] r'[t] - r[t] Phi''[t],
               2
   M Sqrt[r[t] ]                2
  ------------- - r[t] Phi'[t]  + r''[t], 0}
          3
      r[t]
```

The two coupled differential equations of second order in **r** and **Phi** determine the relative motion of the two body problem. The equations for the **r** component contain the total mass as a parameter. To use the two equations in a numerical simulation, we choose a fixed value for the total mass.

```
 parameter = {M->1}

 {M -> 1}
```

Since we plan to solve the equations of motion numerically, we need to supply
initial conditions for the coordinates **r** and **Phi** and their derivatives. To this end,
we define the initial conditions for **r**, **Phi**, **r'**, and **Phi'** as

```
afb = {r[0]==.5,r'[0]==.25,Phi[0]==0,Phi'[0]==1}

{r[0] == 0.5, r'[0] == 0.25, Phi[0] == 0,
   Phi'[0] == 1}
```

Both lists **glr12** and **afb** are joined together with a list of equations and initial
conditions.

```
gli = {glr12[[2]] == 0 /.parameter,
       glr12[[1]] == 0 /.parameter,
       afb}//Flatten

            2
  Sqrt[r[t] ]                    2
{----------- - r[t] Phi'[t]  + r''[t] == 0,
        3
    r[t]
   -2 Phi'[t] r'[t] - r[t] Phi''[t] == 0,
   r[0] == 0.5, r'[0] == 0.25, Phi[0] == 0,
   Phi'[0] == 1}
```

We use this set of equations and initial conditions in the integration process. The
numerical integration of the initial value problem is carried out by the function
NDSolve[ ] which is called by

```
nsol = NDSolve[gli,{r,Phi},{t,0,5},MaxSteps->1000]

{{r -> InterpolatingFunction[{0., 4.90697}, <>],
    Phi -> InterpolatingFunction[{0., 4.90697}, <>]}}
```

The result of **NDSolve[ ]** is an interpolating function for both **r** and **Phi** representing the solution in numerical form. The functions for **r** and **Phi** are defined
in the range $0 < t < 4.90697$ which means that the number of steps given in the
option **MaxSteps** is too small for a complete integration. To use the solutions in a
simple application let us first plot the radial component to demonstrate the periodic
behavior of motion. The function **Plot[ ]** serves to display the radial motion.

```
Plot[Evaluate[r[t]/.nsol],{t,0,4},AxesLabel->{"t","r"}]

-Graphics-
```

We observe from the plot in figure 2.17 that the radial coordinate undergoes an oscillation between a minimal and maximal radius **r**. Thus the relative motion of the two bodies is restricted to a radial domain $rmin < r < rmax$. It is obvious that the maxima are broader than the minima. This means that the particle moves much faster in the perihelion than in the aphelion.

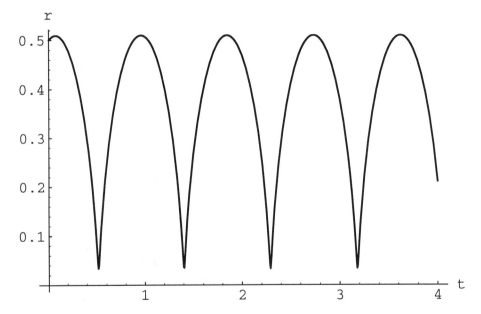

Figure 2.17.   Periodic solution for the radial coordinate.

We next determine the constants of motion of the two body problem. The characteristic of a constant of motion is that it stays the same as time goes by. Later on we will recognize that the knowledge of these invariant quantities allows us to reduce the two-dimensional motion to a one-dimensional motion in an effective potential. The main constants of motion for the two body problem are the total energy and the angular momentum. Since these quantities are constant with time, it is possible to determine their numeric values from the initial conditions. The angular momentum is defined by $r^2\Phi'$. The application of the rule given by the initial conditions **afb** results in the numeric value of the angular momentum

```
L = (r[t]^2 Phi'[t]/.t->0) /. (afb/.Equal->Rule)

0.25
```

By integrating the second equation of **glr12** with respect to time **t**, we are able to derive the total energy. The angular momentum follows if we multiply the first part of **glr12** by **r[t]** and use the function **Integrate[]** to integrate the relation with respect to **t**. We have already determined the numerical value of the angular momentum **L** and can use it in the equation

```
glrr1 = Integrate[glr12[[1]] r[t],t]==L

        2
-(r[t]  Phi'[t]) == 0.25
```

If we solve the resulting equation with respect to **Phi'**, we can use the solution to eliminate the derivative of **Phi** in the second equation of **glr12**. This means that we have eliminated the angular dependence in the second equation and thus decoupled the equations

```
sglr = Solve[glrr1,Phi'[t]]

              -0.25
{{Phi'[t] -> -----}}
               2
             r[t]
```

After replacing **Phi'** and multiplying by **r[t]**, we integrate the second equation of **glr12** to obtain the total energy

```
glrr = Integrate[(glr12[[2]]/. sglr)  r'[t],t] //
          PowerExpand

                       2
 0.03125    M     r'[t]
{-------- - ---- + ------}
      2    r[t]    2
   r[t]
```

The numeric value of the total energy is obtained from the initial conditions by specifying $t = 0$ and using the values of the parameters

```
Energy = (glrr /.t->0) /. parameter /.
        (afb /. Equal->Rule)

{-1.84375}
```

From the energy we can extract a potential which depends only upon **r**. Neglecting the changes in time, we express the so–called effective potential in the form

```
Ueff = -M/r+L^2/(2 r^2) /. parameter

0.03125    1
-------  - -
   2       r
  r
```

A graphical representation of this potential is given by

```
pl2 = Plot[Evaluate[{Ueff /. parameter,Energy}],{r,0,3},
     PlotRange->{-8,2},
     PlotStyle->{RGBColor[1,0,0],
                 RGBColor[0,1,0]},
     AxesLabel->{"r","Ueff"}]

-Graphics-
```

In figure 2.18, we have plotted the total energy value to signify the energy level on which the motion of the relative particle in the potential takes place. This energy level locates the two points at which the particle comes to rest and reverses its direction. Between these two turning points the particle gains kinetic energy which means that the particle velocity increases. To see the radial motion, we make a small animation of the motion demonstrating the movement in the radial coordinate. We create a table containing 20 time steps, each of which represents a different location of the particle in the potential. To graphically represent the location of the particle in the potential, we use **Point**[ ] and the color function **RGBColor**[ ].

```
tab1 = Table[{RGBColor[0,0,1],
             Point[{r[t]/.nsol,Egs]//Flatten]},
        {t,0,4,.4}];
```

The different pictures of the particle are superimposed on the plot of the potential. The sequence created in this way is shown by **Show**[ ] and a **Do**[ ] loop

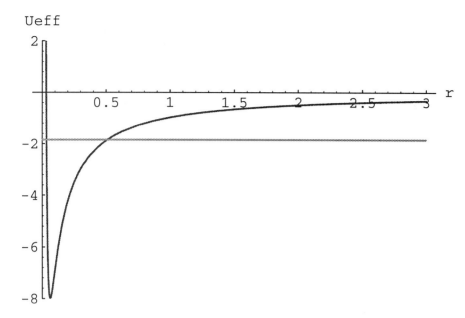

Figure 2.18. The intersection between the total energy and the effective potential. The motion in the potential takes place between the two intersecting points.

```
pl1 = {}

{}

Do[AppendTo[pl1,Show[{pl2,Graphics[{PointSize[0.02],tab1[[i]]}]}] ],
{i,1,Length[tab1]}]

pl2 = Partition[pl1,2];

Show[GraphicsArray[pl2]]

-GraphicsArray-
```

GraphicsArray[ ] yields a printed representation of the animation (figure 2.19). The numerical integration provides the solution of the two particle problem for a certain time domain in radial and angular components. The question now is how can we use this information to represent the relative real space motion. If we use the solutions **nsol** and apply them to the transformation rules **s6**, we obtain the solution in relative coordinates

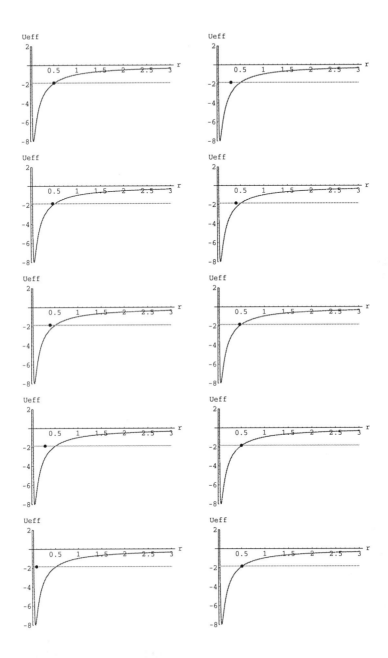

Figure 2.19.   Animation of the motion in the effective potential. The particle in the potential moves between the two turning points.

```
s7 = s6  /. nsol

{{x -> Function[t, InterpolatingFunction[{0.,
        4.90697}, <>][t]
      Cos[InterpolatingFunction[{0., 4.90697},
        <>][t]]], y ->
    Function[t, InterpolatingFunction[{0.,
        4.90697}, <>][t]
      Sin[InterpolatingFunction[{0., 4.90697},
        <>][t]]], z -> Function[t, 0]}}
```

A graphical representation of the solution using **ParametricPlot**[] indicates that the orbit in relative coordinates is an ellipse as shown in figure 2.20.

```
pl3 = ParametricPlot[{x[t],y[t]}/.s7,{t,0,4}]

-Graphics-
```

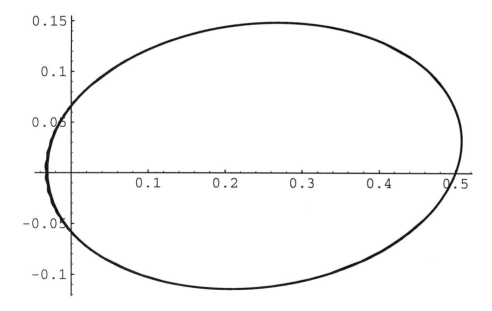

Figure 2.20.   Representation of the orbit in relative coordinates.

Using the solution in relative coordinates, we can show the relative motion of the particle on this track. To animate the motion, we again create a table containing several locations of a particle along this path. Matching the pictures of the points with the picture of the elliptic track, we create an animation showing the motion of the relative particle in space (see figure 2.21).

```
tab2 = Table[{RGBColor[0,0,1],Point[{x[t],y[t]}/.
      s7//Flatten]},
      {t,0,4,.4}];

plh3 = {}

{}

Do[AppendTo[plh3,
    Show[{pl3,Graphics[{PointSize[0.02],tab2[[i]]}]}],
  AspectRatio->Automatic ] ],
{i,1,Length[tab2]}]

plh3 = Partition[plh3,2]

{{-Graphics-, -Graphics-},
  {-Graphics-, -Graphics-},
  {-Graphics-, -Graphics-},
  {-Graphics-, -Graphics-},
  {-Graphics-, -Graphics-}}
```

An abridged version of the movement is printed with the help of a graphics array.

```
Show[GraphicsArray[plh3]]

-GraphicsArray-
```

We have demonstrated the animation of motion in real coordinates. The discussion has shown that the motion is a combination of the center of mass motion and the relative motion. To simplify the graphical representation in real coordinates, we assumed that the center of mass is at rest. This condition is expressed by the rule

```
s8 = {R1->Function[t,0],
      R2->Function[t,0],
      R3->Function[t,0]}

{R1 -> Function[t, 0], R2 -> Function[t, 0],
  R3 -> Function[t, 0]}
```

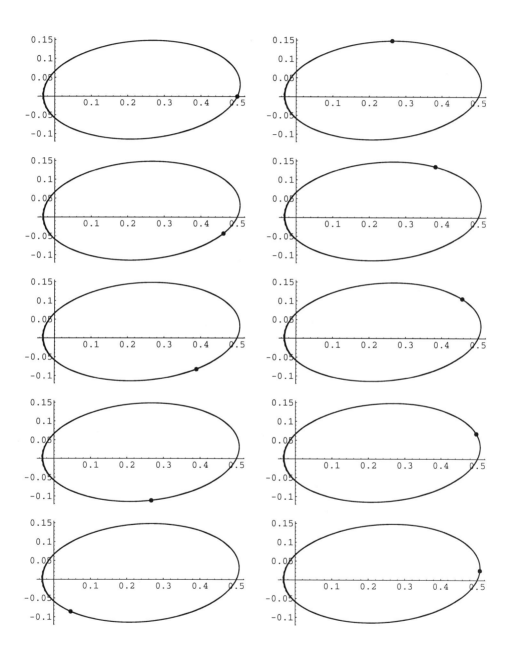

Figure 2.21.   Relative elliptic motion.

For a real representation of the motion, we either use mass **m1** or **m2** as an input parameter. We can input information about **m1** by means of the **Input[ ]** function

```
m1 = Input["mass m1: "]

0.4
```

The second mass is determined by conservation of the total mass which we can write in the form

```
m2 = M - m1 /. parameter

0.6
```

The original coordinates in real space are represented by the two–dimensional vectors **r1** and **r2**

```
r1 = m2 {x[t],y[t]}/(m1 + m2) /. s7;
```

and

```
r2 = -m1 {x[t],y[t]}/(m1 + m2) /. s7;
```

These two sets of coordinates are graphically presented by the function **ParametricPlot[ ]**. The result is a picture containing two ellipses which overlap each other and share the common center of mass as shown in figure 2.22.

```
pl4 = ParametricPlot[Evaluate[{r1,r2}],{t,0,4}]

-Graphics-
```

To illustrate the real motion of the two particles is possible by applying the superposition technique introduced above. We create two tables containing the positions of the bodies at certain times. Using the function **Show[ ]**, we graphically represent the orbits and the positions of the bodies in a sequence of plots.

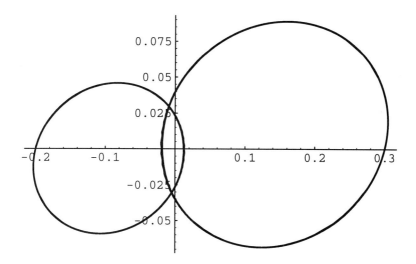

Figure 2.22.   Real paths of the two bodies.

```
tab3 = Table[{RGBColor[1,0,0],Point[r1//Flatten],
            Line[{{0,0},r1[[1]]}]},
      {t,0,4,.4}];

tab4 = Table[{RGBColor[0,1/2,0],Point[r2//Flatten],
            Line[{{0,0},r2[[1]]}]},
      {t,0,4,.4}];

plh4 = {}

{}

Do[AppendTo[plh4,
   Show[{pl4,Graphics[{PointSize[0.02],tab3[[i]]}],
            Graphics[{PointSize[0.03],tab4[[i]]}]}]],
{i,1,Length[tab4]}]

plh4 = Partition[plh4,2]

{{-Graphics-, -Graphics-},
  {-Graphics-, -Graphics-},
  {-Graphics-, -Graphics-},
  {-Graphics-, -Graphics-},
  {-Graphics-, -Graphics-}}

Show[GraphicsArray[plh4]]

 -GraphicsArray-
```

The animation of the motion is again printed with **GraphicsArray[ ]**, as shown in figure 2.23.

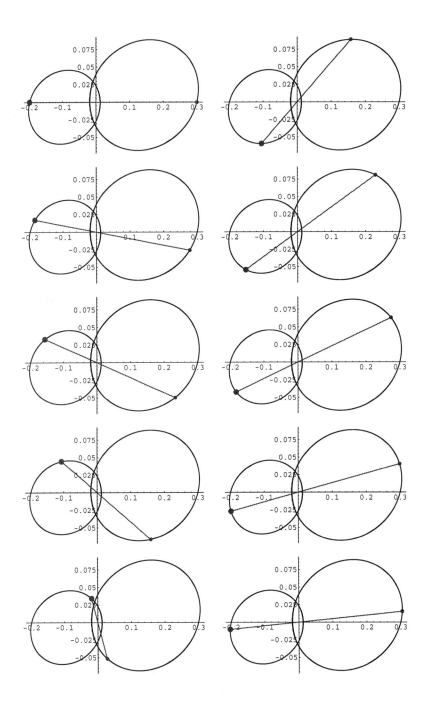

Figure 2.23.   Animation of the motion in real coordinates.

We have solved the two body problem without defining any *Mathematica* functions. All calculations of this section are done using a *Mathematica* notebook. The notebook is available on the accompanying disk under the name *kepler.ma*. You can use this notebook to produce the animations discussed above or to examine the two body motion for other conditions. Experimenting with the notebook will show you among other things some aspects of the planetary motion within Newton's theory.

2.7 Vibrations of a membrane

A membrane is a three–dimensional elastic body whose thickness is much smaller than its lateral extension. The equilibrium state of the membrane is characterized by equal tensions $p = p_{xx} = p_{yy}$ in the (x, y) plane. The simple vibrations of a membrane are those oscillations which are much smaller in their amplitude than the lateral extension of the membrane. By this definition we exclude nonlinear phenomena connected with large amplitudes of the membrane. The assumption of small amplitudes has the consequence that we only consider forces directed perpendicular to the membrane along the z axis [2.2]. The vertical forces are used to derive the equation of motion for an infinitesimal area $dxdy$. We assume that the forces act on the boundaries of the area $dxdy$ in the x and y directions perpendicular to the boundary lines. These forces in the plane shrink and shear the area in a specific way. For small amplitudes the z component of the force acting on the membrane is

$$qp\frac{\partial^2 \zeta(x, y, t)}{\partial x^2}dx = hp\frac{\partial^2 \zeta(x, y, t)}{\partial x^2}dxdy. \tag{2.68}$$

The constants in equation (2.68) denote the tension p and the area of a membrane element $q = hdy$ where h denotes the extension in x direction. The variable ζ denotes the vertical deviation of the membrane. The reasoning used to explain the behavior of the forces is also applicable to the remaining sides of the area element $dxdy$. The forces on the side elements are given by

$$hp\left(\frac{\partial^2 \zeta}{\partial x^2} + \frac{\partial^2 \zeta}{\partial y^2}\right)dxdy. \tag{2.69}$$

According to Newton's law, we know that the acting forces are equivalent to the product of the mass $\varrho hdxdy$ and the acceleration $\partial^2 \zeta/\partial t^2$, which results in the equation of motion for the transversal vibrating membrane

$$\frac{\partial^2 \zeta}{\partial t^2} = c^2\left(\frac{\partial^2 \zeta}{\partial x^2} + \frac{\partial^2 \zeta}{\partial y^2}\right) \tag{2.70}$$

where $c = (p/\varrho)^{1/2}$ is the velocity of the waves. Since equation (2.70) is a second order partial differential equation, we must specify the initial and boundary conditions under which the vibration takes place in order to obtain a unique solution.

As an example we will study a circular membrane with fixed boundaries. The transformation of the plane Cartesian coordinates in the (x, y) plane to the polar coordinates $x = \varrho \cos \phi$ and $y = \varrho \sin \phi$ yields the equation of motion in the form

$$\frac{\partial^2 \zeta}{\partial t^2} = c^2 \left\{ \frac{1}{\varrho} \frac{\partial}{\partial \varrho} \left(\varrho \frac{\partial \zeta}{\partial \varrho} \right) + \frac{1}{\varrho^2} \frac{\partial^2 \zeta}{\partial \phi^2} \right\}. \tag{2.71}$$

If we fix the membrane on the circular boundary, we have the boundary condition in the form

$$\zeta(\varrho = R, \phi, t) = 0. \tag{2.72}$$

By an appropriate scaling of the wave velocity c, we can always reduce the maximum radius of the membrane to the value $R = 1$. Consequently the boundary condition (2.72) reduces to

$$\zeta(\varrho = 1, \phi, t) = 0. \tag{2.73}$$

One procedure to solve the boundary problem of a circular membrane is the method of eigenfunction expansion described by Courant & Hilbert [2.4]. This procedure assumes that the vertical deviation $\zeta(\varrho, \phi, t)$ is separable into a product of a time–dependent part and a spatial part given by

$$\zeta(\varrho, \phi, t) = \Gamma(\varrho, \phi) \Delta(t). \tag{2.74}$$

By applying the expression (2.74) to equation (2.71), we can separate the time–dependent parts from the spatial parts of the vertical deviation

$$\frac{d^2 \Delta(t)}{dt^2} \frac{1}{\Delta} = c^2 \frac{1}{\Gamma(\varrho, \phi)} \left\{ \frac{1}{\varrho} \frac{\partial}{\partial \varrho} \left(\varrho \frac{\partial \Gamma}{\partial \varrho} \right) + \frac{1}{\varrho^2} \frac{\partial^2 \Gamma}{\partial \phi^2} \right\} = -\omega^2. \tag{2.75}$$

Because the time–dependent and spatial components in (2.75) are separated, we obtain the relations

$$\frac{d^2 \Delta}{dt^2} = -\omega^2 \Delta \tag{2.76}$$

$$-k^2 \Gamma = \left\{ \frac{1}{\varrho} \frac{\partial}{\partial \varrho} \left(\varrho \frac{\partial \Gamma}{\partial \varrho} \right) + \frac{1}{\varrho^2} \frac{\partial^2 \Gamma}{\partial \phi^2} \right\} \tag{2.77}$$

where k is the wave number. Equation (2.76) describes the harmonic part of the time domain, and relation (2.77) determines the spatial variations of the amplitude. The dispersion relation connecting the frequency of the wave with the wave vector is $\omega = ck$.

The amplitude equation (2.77) is separated by the same arguments as above. The separation is given by $\Gamma(\varrho, \phi) = h(\varrho)g(\phi)$. The constant of separation is now denoted by m^2. Using the separation expression for Γ in equation (2.77), we get

$$\frac{\varrho^2 h''}{h} + \frac{\varrho h'}{h} + k^2 \varrho^2 = -\frac{g''}{g} = m^2. \tag{2.78}$$

Primes denote the differentiation with respect to the independent coordinate. The second separation finally yields

$$g'' + m^2 g \;=\; 0 \tag{2.79}$$

$$h'' + \frac{1}{\varrho}h' + \left(k^2 - \frac{m^2}{\varrho^2}\right)h \;=\; 0. \tag{2.80}$$

Equation (2.79) is again a second order equation describing harmonic oscillation but this time for the angular coordinate. The general solution for equation (2.79) is given by

$$g(\phi) = A\sin(m\phi) + B\cos(m\phi). \tag{2.81}$$

The second equation for the spatial components equation (2.80) is reduced to the standard form of Bessel's differential equation if we introduce the transformation $x = k\varrho$. Bessel's equation reads

$$h''(x) + \frac{1}{x}h'(x) + \left(1 - \frac{m^2}{x^2}\right)h(x) = 0. \tag{2.82}$$

The solution of the radial equation (2.82) for the vibrating membrane is

$$h(x) = J_m(x) = J_m(k\varrho) \qquad \text{with} \qquad m = 0, 1, 2, \ldots \tag{2.83}$$

where $J_m(x)$ is a Bessel function of the first kind, finite at $x = 0$, which is denoted **BesselJ[m,z]** in *Mathematica*. The spatial amplitude Γ is represented by

$$\Gamma_m(\varrho, \phi) = J_m(k\varrho)\left\{A\sin(m\phi) + B\cos(m\phi)\right\}. \tag{2.84}$$

The subscript m of Γ_m indicates that different solutions exist in the entire solution process. Up to now we have not included the boundary condition in our calculations. The boundary condition (2.73) is also separated and reads

$$\begin{aligned} \Gamma(\varrho = 1, \phi) &\;=\; h(\varrho = 1)g(\phi) = 0 \\ J_m(k) &\;=\; 0. \end{aligned} \tag{2.85}$$

Thus we have reduced the determination of the eigenvalues k to finding the roots of the m-th Bessel function. It is well known that the Bessel function J_m possesses infinitely many roots [1.2]; i.e., we have to determine for each m an infinite number of eigenvalues k from equation (2.85). It is convenient to introduce a special numbering for the eigenvalues k in the form of $k = k_{m,n}$. The numbering fixes the order of the Bessel function by m and the n-th root of the Bessel function J_m by n. One consequence of this numbering is shown for the dispersion relation

$$\begin{aligned} \omega &\;=\; ck \tag{2.86} \\ &\;=\; ck_{m,n} = \omega_{m,n}. \tag{2.87} \end{aligned}$$

The time–dependent solution of equation (2.71) for fixed m and n values is thus given by

$$
\begin{aligned}
\zeta_{m,n}(\varrho,\phi,t) \;=\; & J_m(k_{m,n}\varrho)\left\{A_{m,n}\sin(m\phi)+B_{m,n}\cos(m\phi)\right\} \\
& * \left\{D_{m,n}\sin(\omega_{m,n}t)+E_{m,n}\cos(\omega_{m,n}t)\right\}.
\end{aligned}
\tag{2.88}
$$

Since equation (2.71) is a linear equation of motion the total solution is a superposition of the particular solutions (2.88). Summing over all representations (2.88) yields

$$
\zeta(\varrho,\phi,t)=\sum_{m,n=0}^{\infty} C_{m,n}\zeta_{m,n}(\varrho,\phi,t).
\tag{2.89}
$$

The explicit representation of this series is

$$
\begin{aligned}
\zeta(\varrho,\phi,t) \;=\; & \sum_{m,n=0}^{\infty} \tilde{c}_{m,n}J_m(k_{m,n}\varrho)\sin(m\phi)\left\{D_{m,n}\sin(\omega_{m,n}t)+E_{m,n}\cos(\omega_{m,n}t)\right\} \\
& +\sum_{m,n=0}^{\infty} \tilde{d}_{m,n}J_m(k_{m,n}\varrho)\cos(m\phi)\left\{D_{m,n}\sin(\omega_{m,n}t)+E_{m,n}\cos(\omega_{m,n}t)\right\}
\end{aligned}
\tag{2.90}
$$

where $\tilde{c}_{m,n} = C_{m,n}A_{m,n}$ and $\tilde{d}_{m,n} = C_{m,n}B_{m,n}$. The unknown quantities in the expansion (2.90) are the time–independent coefficients $\tilde{c}_{m,n}D_{m,n}$, $\tilde{c}_{m,n}E_{m,n}$, $\tilde{d}_{m,n}D_{m,n}$ and $\tilde{d}_{m,n}E_{m,n}$. At this point of the solution procedure, the initial conditions $\zeta(\varrho,\phi,t=0)$ and $\partial\zeta(\varrho,\phi,t=0)/\partial t$ come into play. Since the expansion coefficients are independent of time, we can use the orthogonality of the Bessel functions to calculate the expansion coefficients. If, in addition, we assume that the initial conditions given by $\zeta(\varrho,\phi,0)=f_1(\varrho,\phi)$ and $\zeta'(\varrho,\phi,0)=f_2(\varrho,\phi)$ are separable in the form $\zeta(\varrho,\phi,0)=h_1(\varrho)g_1(\phi)$ and $\zeta'(\varrho,\phi,0)=h_2(\varrho)g_1(\phi)+h_1(\varrho)g_2(\phi)$, the coefficients are represented by the integrals

$$
\tilde{c}_{m,n}D_{m,n} = \frac{1}{\omega_{m,n}\sqrt{\pi J_{m,n}}}\left\{I_2^{m,n}I_3^{m,n}+I_1^{m,n}I_4^{m,n}\right\}
\tag{2.91}
$$

$$
\tilde{d}_{m,n}D_{m,n} = \frac{1}{\omega_{m,n}\sqrt{\pi J_{m,n}}}\left\{I_2^{m,n}I_5^{m,n}+I_1^{m,n}I_6^{m,n}\right\}
\tag{2.92}
$$

$$
\tilde{c}_{m,n}E_{m,n} = \frac{1}{\sqrt{\pi J_{m,n}}}I_1^{m,n}I_3^{m,n}
\tag{2.93}
$$

$$
\tilde{d}_{m,n}E_{m,n} = \frac{1}{\sqrt{\pi J_{m,n}}}I_1^{m,n}I_5^{m,n}
\tag{2.94}
$$

$$
\tag{2.95}
$$

where

$$J_{m,n} = \int_0^1 J_m^2(k_{m,n}, \varrho)\varrho d\varrho \qquad (2.96)$$

$$I_1^{m,n} = \int_0^1 h_1(\varrho) J_m(k_{m,n}\varrho)\varrho d\varrho \qquad (2.97)$$

$$I_2^{m,n} = \int_0^1 h_2(\varrho) J_m(k_{m,n}\varrho)\varrho d\varrho \qquad (2.98)$$

$$I_3^{m,n} = \int_0^{2\pi} g_1(\phi)\sin(m\phi)d\phi \qquad (2.99)$$

$$I_4^{m,n} = \int_0^{2\pi} g_2(\phi)\sin(m\phi)d\phi \qquad (2.100)$$

$$I_5^{m,n} = \int_0^{2\pi} g_1(\phi)\cos(m\phi)d\phi \qquad (2.101)$$

$$I_6^{m,n} = \int_0^{2\pi} g_2(\phi)\cos(m\phi)d\phi \qquad (2.102)$$

where $\omega_{m,n} = ck_{m,n}$. Once the expansion coefficients are known, the solution in form (2.90) is also known. One essential step of this procedure is the determination of the roots of the Bessel function and the calculation of the integrals $I_i^{m,n}$. It is likely that the infinite sum (2.90) cannot be computed by *Mathematica* since the involved expressions are too complicated. Therefore we need to use the infinite series at finite order. This kind of formulation will cause an error in our calculation which can be kept small if the approximation order is chosen sufficiently large.

The solution procedure for the circular membrane is collected in the package **Membrane'** listed below.

```
––––––––––––––––––––––––– File membrane.m –––––––––––––––––––

BeginPackage["Membrane'"]

Clear[fRoot, listeN, BesselRoot, fSort, Evolution, TimeSeries]

Evolution::usage = "Evolution[h1_,h2_,g1_,g2_,t1_,order_,number_]
calculates the vibration of a membrane for the initial conditions
h1, h2, g1 and g2 at time t1. The input variable order determines the order
of the Bessel function while number determines the number of roots of the
Bessel function. The solution is presented by a surface plot in three
dimensions."

TimeSeries::usage ="TimeSeries[t_] plots a sequence of states of the vibrating
membrane. This function uses the analytic expression derived by the function
Evolution. The function Evolution[] is used prior to using TimeSeries[]."


Begin["'Private'"]

(* --- Determine the roots of the Bessel functions.
       Needed for Mathematica version 2.0 and smaller.
       For later versions of Mathematica use the standard
       package BesselZeros .                            --- *)
```

```
fRoot[n_] := Block[{rootStart, listRoot, root},
(* --- initial values for the numerical calculation of the roots --- *)
   listRoot = {0, 1, 2, 2, 3, 5, 6, 8, 8, 9, 10, 11, 12, 13, 14, 15,
        17, 18, 19, 20, 21};
    rootStart = listRoot[[n + 1]];
    root = x /. FindRoot[BesselJ[n, x] == 0, {x, rootStart + 3}];
    root]

fRoot[n_, inRoot_] :=
  Block[{startRoot,root}, startRoot = inRoot;
    root = x /. FindRoot[BesselJ[n, x] == 0, {x, startRoot + 3}];
    root]

(* --- create a list of roots --- *)

listeN[n_,nn_]:=Block[{liste,root},
            root = fRoot[n];
            liste = {root};
            Do[AppendTo[liste,root = fRoot[n,root]],{k,0,nn}];
            liste]

BesselRoot[ordnung_,zahl_]:=Block[{listR, sublist},
            listR = {};
            Do[AppendTo[listR,sublist = listeN[k,zahl]],{k,0,ordnung}];
            listR]

fSort[aMN_List]:=Block[{helpList,aM},
            helpList = {};
            aM = N[Flatten[aMN]];
            len = Length[aM];
            Do[If[Abs[Part[aM,m]] > 10^(-8),
                AppendTo[helpList,Part[aM,m]],
                AppendTo[helpList,0]],{m,1,len}];
            helpList]

(* --- graphical representation of the vibration --- *)

Evolution[h1_,h2_,g1_,g2_,t1_,order_,number_]:=
            Block[{lbMNcMN, lbMNdMN, laMNcMN, laMNdMN,
                listNorm, l1MN, l2MN, l3MN, l4MN, l5MN, l6MN,
                istEigenval, listEntwick, listLen, sublistLen, c,
                h1h, h2h, g1h, g2h, x1, aMNcMN = {}, aMNdMN = {},
                bMNcMN = {}, bMNdMN = {}, i1MN = {}, i2MN = {},
                i3MN = {}, i4MN = {}, i5MN = {}, i6MN = {},
                listNorm = {}
        },

            h1h = h1 /. f_[x1_]->f[r];
            h2h = h2 /. f_[x1_]->f[r];
            g1h = g1 /. f_[x1_]->f[x];
            g2h = g2 /. f_[x1_]->f[x];

(* --- set propagation velocity to c=1   --- *)
            c = 1;
(* --- determine the roots of the Bessel function --- *)
            Off[FindRoot::frmp];
```

```
                    listEigenval = BesselRoot[order,number-1];
                    On[FindRoot::frmp];
                    listLen = Length[listEigenval];
                    sublistLen = Length[Part[listEigenval,1]];
 (* --- normalization of the Bessel function --- *)
      Off[NIntegrate::ploss];

             Do[
             Do[AppendTo[listNorm,
                  NIntegrate[r*
                   BesselJ[k-1,Part[Part[listEigenval,k],n]*r]*
                   BesselJ[k-1,Part[Part[listEigenval,k],n]*r]
                               ,{r,0,1}]],
                               {n,1,sublistLen}],
                               {k,1,listLen}];

                     listNorm = Sqrt[listNorm];
                     listNorm = N[Partition[listNorm,sublistLen]];
 (* --- determine the expansion coefficients --- *)
             Do[
             Do[AppendTo[i1MN,
                  NIntegrate[(h1h*r*
                  BesselJ[k-1,Part[Part[listEigenval,k],n]*r]),{r,0,1}]],
                               {n,1,sublistLen}],
                               {k,1,listLen}];
                 i1MN = N[Partition[i1MN,sublistLen]];

             Do[
             Do[AppendTo[i2MN,
                  NIntegrate[(h2h*r*
                  BesselJ[k-1,Part[Part[listEigenval,k],n]*r]),{r,0,1}]],
                               {n,1,sublistLen}],
                               {k,1,listLen}];
                 i2MN = N[Partition[i2MN,sublistLen]];

             Do[
             Do[AppendTo[i3MN,
                  NIntegrate[g1h*Sin[(k-1)*x],{x,0,2Pi}]],
                               {n,1,sublistLen}],
                               {k,1,listLen}];
                 i3MN = N[Partition[i3MN,sublistLen]];

             Do[
             Do[AppendTo[i4MN,
                  NIntegrate[g2h*Sin[(k-1)*x],{x,0,2Pi}]],
                               {n,1,sublistLen}],
                               {k,1,listLen}];
                 i4MN = N[Partition[i4MN,sublistLen]];

             Do[
             Do[AppendTo[i5MN,
                  NIntegrate[g1h*Cos[(k-1)*x],{x,0,2Pi}]],
                               {n,1,sublistLen}],
                               {k,1,listLen}];
                 i5MN = N[Partition[i5MN,sublistLen]];

             Do[
```

```
        Do[AppendTo[i6MN,
            NIntegrate[g2h*Cos[(k-1)*x],{x,0,2Pi}]],
                        {n,1,sublistLen}],
                        {k,1,listLen}];
        i6MN = N[Partition[i6MN,sublistLen]];

    On[NIntegrate::ploss];

        Do[
        Do[AppendTo[aMNcMN,
            (Part[Part[i2MN,k],n]*Part[Part[i3MN,k],n]+
            Part[Part[i1MN,k],n]*Part[Part[i4MN,k],n])/
    (Part[Part[listNorm,k],n]*c*Part[Part[listEigenval,k],n]*
                        Sqrt[Pi]) ],
                        {n,1,sublistLen}],
                        {k,1,listLen}];

        aMNcMN = fSort[aMNcMN];
        aMNcMN = Partition[aMNcMN,sublistLen];

        Do[
        Do[AppendTo[aMNdMN,
            (Part[Part[i2MN,k],n]*Part[Part[i5MN,k],n]+
            Part[Part[i1MN,k],n]*Part[Part[i6MN,k],n])/
    (Part[Part[listNorm,k],n]*c*Part[Part[listEigenval,k],n]*
                        Sqrt[Pi])],
                        {n,1,sublistLen}],
                        {k,1,listLen}];

        aMNdMN = fSort[aMNdMN];
        aMNdMN = Partition[aMNdMN,sublistLen];

        Do[
        Do[AppendTo[bMNcMN,
            (Part[Part[i1MN,k],n]*Part[Part[i3MN,k],n])/
            (Part[Part[listNorm,k],n]*Sqrt[Pi])
                        ],
                        {n,1,sublistLen}],
                        {k,1,listLen}];

        bMNcMN = fSort[bMNcMN];
        bMNcMN = Partition[bMNcMN,sublistLen];

        Do[
        Do[AppendTo[bMNdMN,
            (Part[Part[i1MN,k],n]*Part[Part[i5MN,k],n])/
            (Part[Part[listNorm,k],n]*Sqrt[Pi])
                        ],
                        {n,1,sublistLen}],
                        {k,1,listLen}];
        bMNdMN = fSort[bMNdMN];
        bMNdMN = Partition[bMNdMN,sublistLen];

(* --- calculate the sum --- *)

phi1 =
Sum[
```

```
     BesselJ[k-1,Part[Part[listEigenval,k],n]*r]*Sin[(k-1)*y]*
     (Part[Part[aMNcMN,k],n]*Sin[Part[Part[listEigenval,k],n]*c*t] +
     Part[Part[bMNcMN,k],n]*Cos[Part[Part[listEigenval,k],n]*c*t]),
                         {n,1,sublistLen},{k,1,listLen}] +
Sum[
     BesselJ[k-1,Part[Part[listEigenval,k],n]*r]*Cos[(k-1)*y]*
     (Part[Part[aMNdMN,k],n]*Sin[Part[Part[listEigenval,k],n]*c*t] +
     Part[Part[bMNdMN,k],n]*Cos[Part[Part[listEigenval,k],n]*c*t]),
                         {n,1,sublistLen},{k,1,listLen}];
     phi = phi1 /. t->t1;
(* --- create a plot --- *)
ParametricPlot3D[{r*Cos[y],r*Sin[y],phi},{r,0,1},{y,0,2Pi}]
                         ]

(* --- show a time series of pictures --- *)

TimeSeries[t1_] := Block[{phi}, phi = phi1 /. t -> t1;
    ParametricPlot3D[{r*Cos[y], r*Sin[y], phi}, {r,0,1}, {y,0,2*Pi}]]
End[]
EndPackage[]
```

x The results of a calculation for $h_1(\varrho) = \cos(\varrho)$, $h_2 = 1$, $g_1(\phi) = \sin(\phi)$, and $g_2 = 1$ for $m = 5$, and $n = 5$ are shown in figures 2.24–2.27. The function used in creating the pictures takes the sequence **Evolution[Cos[x],1,Sin[x],1,0.1,5,5]**.

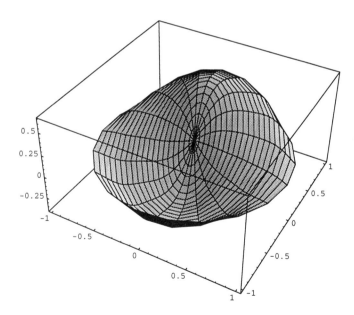

Figure 2.24.   Vibration of a membrane under the conditions $h_1(\varrho) = \cos(\varrho)$, $h_2(\varrho) = 1.0$, $g_1(\phi) = \sin(\phi)$, and $g_2(\phi) = 1.0$. The time is $t = 0.1$.

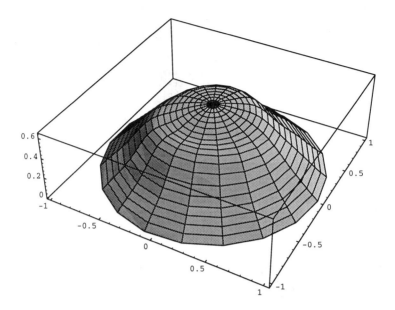

Figure 2.25. Time is $t = 0.5$.

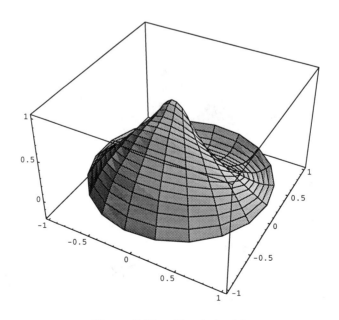

Figure 2.26. Time is $t = 1.0$.

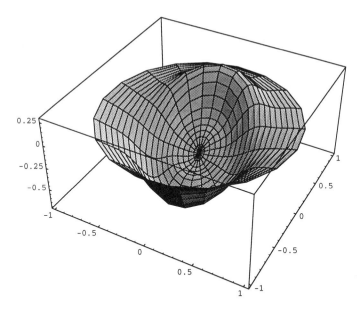

Figure 2.27. Time is $t = 1.5$.

2.8 Dynamic formulation

In this section we discuss the dynamic formulation for equations of motion in a more general context. In classical mechanics, we distinguish between two main types of equivalent dynamic formulations:

1. the Lagrange formulation

2. the Hamiltonian formulation.

Both formulations are equivalent and describe the motion of a system in two different ways. We derived the canonical form of the Hamilton formulation by means of the Poisson bracket at the beginning of this chapter. In this section we show that the Lagrange formalism delivers equations of motion which are identical with Newton's second law. We also consider the dynamic formulation of point and continuous systems.

2.8.1 Lagrange formulation

The Lagrange formulation of a dynamic system is based on Hamilton's principle. This principle states that the action of a system must be an extremum. In this case the action is a quantity with dimension energy $\times$ time. The principle is simply stated as an assumption for which no proof exists. Therefore, the analytic form of the action function determines the dynamics of the entire system.

To write down an explicit expression for the action, we have to make a number of assumptions about the energetic properties of the system. These assumptions are in close analogy to the axioms established by Newton's theory and cannot be derived from fundamental principles. To state the contents of the Lagrange formulation, we assume that the system under consideration is characterized by the Lagrangian

$$L = L(q_1, q_2, q_3, ..., q_f; q'_1, q'_2, q'_3, ..., q'_f) \qquad (2.103)$$

where $q_1, q_2, \ldots, q_f$ and $q'_1, q'_2, \ldots, q'_f$ are the f generalized coordinates and velocities of the system. The explicit form of the Lagrangian follows from physical considerations of the energetic properties of the system. Let us collect the coordinates and velocities in a set of variables $q = q_1, q_2, q_3, \ldots, q_f$ and q', allowing the short–hand notation $L = L(q, q')$. Hamilton's principle of least action can then be formulated as follows:

The motion of the system in the time interval $t_1 \le t \le t_2$ is governed by the extremum of the integral

$$S = \int_{t_1}^{t_2} L(q, q', t) dt. \qquad (2.104)$$

In the literature, the integral S is sometimes called a functional, which is often used in the application of field theory. L denotes the functional density of the action S. In general, this density can also depend on higher derivatives with respect to time or spatial coordinates. The objective is to determine the extremum of the action. By an extremum we mean the state in which S has its minimum or maximum value for a given function q. To find the extremum of S from a huge number of states, we have to vary function q. The variations of q show us whether S increases or decreases. To check the action values S for different forms of q, we need to introduce a test function w, which changes the properties of S. This test function depends on a parameter e and allows us to control the influence of the variations. The variation of q is assumed to be

$$q(t; e) = q(t; 0) + ew(t). \qquad (2.105)$$

In case of $e = 0$, the variation reduces to q. The derivation of the equations of motion according to the principle of least action is primarily based on the calculus of variations.

The calculus of variations was first invented by the Bernoulli brothers around 1696. Their main assumption is that there exists a generating function of motion depending on a parameter determining the specific value of action S. S is varied by replacing q with $q + ew$. After replacing q and all of its higher derivatives in S, we have to determine the extreme value of S. Consequently S has become a function of e. The maximum or minimum of S is obtained by using the standard procedure of calculus for finding extrema. We need to calculate the derivative of S with respect to e

$$\frac{dS(e)}{de}\bigg|_{e=0} = 0. \qquad (2.106)$$

We now consider some examples of variational problems.

2.8.1.1 Shortest distance in the plane

In Cartesian coordinates the infinitesimal distance between two points is given by the line element

```
ds = Sqrt[dx^2 + dy^2];
```

or expressed in a slightly different way by

```
ds = Sqrt[1 +D[y[x],x]^2] dx;
```

The total length between the two points is calculated by integrating along the path connecting the two points p_1 and p_2

```
s = Integrate[ds/dx,{x,x1,x2}];
```

However, for a plane in Euclidian space we know that the shortest distance between two points is a straight line represented by

```
yh = a x + b

b + a x
```

where **a** and **b** are the slope of the line and the intersection of the line with the vertical axis of the coordinate system respectively. Our aim is to show that action S assumes a minimum value for the solution **yh**. In order to prove this assumption we consider an x–interval ranging from $0 \leq x \leq 2\pi$. Modifying **yh** by the test function **w**

```
w[x] = Sin[4x]

Sin[4 x]
```

If we choose $a = 1$ and $b = 0$ in **yh**, we can write the entire function in the form

```
rule1 = y-> Function[{x,e},x+e Sin[4x]]

y -> Function[{x, e}, x + e Sin[4 x]]
```

The expression contained in **rule1** is a *Mathematica* rule. We use a pure function definition for replacing **y** as well as its derivatives. The integrand of action S is given by

```
integ = Sqrt[1+D[y[x,e]/.rule1,x]^2]

                     2
Sqrt[1 + (1 + 4 e Cos[4 x]) ]
```

We define S by numerical integration using the variable **integ** in the definition of the function **S[]**

```
S[e1_]:=NIntegrate[integ /. e->e1,{x,0,2Pi}]
```

The question is whether S takes a minimum or a maximum for $e = 0$. To check the behavior of the action, we simply plot S as a function of parameter e. Using the built–in function **Plot[]** for the graphical representation of the action we get for the interval $-1 \leq e \leq 1$ (see figure 2.28).

```
Plot[Evaluate[S[e1]],{e1,-1,1},AxesLabel->{"e","S"}]

-Graphics-
```

We also plot **y** and a couple of variations to get a feeling for the behavior of **y** in different states. Different values of e are shown in figure 2.29.

```
Plot[Evaluate[{y[x,1],y[x,1/2]} /. rule1],{x,0,2Pi}]

-Graphics-
```

We observe in figure 2.29 that the test function given by the expression

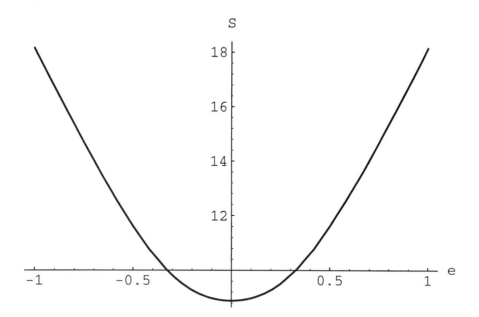

Figure 2.28.   Minimum of S for the shortest distance between two points.

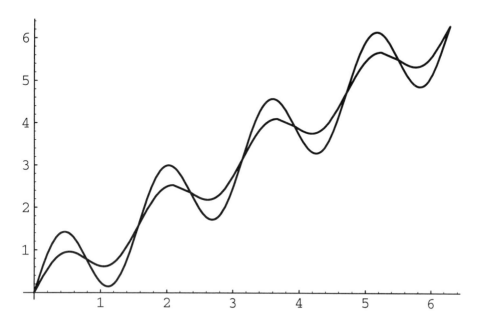

Figure 2.29.   Variations of the function $y(x; e)$ for two values of $e = 1/2$ and $e = 1$.

```
w[x] = e Sin[4 x]
```

vanishes at the endpoints of the interval. The property of the test function vanishing at the boundaries is known as the fundamental lemma of the calculus of variation.

2.8.1.2 The Brachistochrone problem The Brachistochrone problem was first solved by Johann Bernoulli in 1696. The Brachistochrone problem considers a particle sliding along a path between two points in a homogeneous force field. The distance between the two points must be tracked in the shortest time possible. For a simple mathematical formulation we choose the origin of the coordinate system as the starting point of the motion. We assume that the homogeneous force field is directed along the negative x axis (see figure 2.30).

```
pl1 = Plot[-x + Sin[4 x],{x,0,2Pi},
        AxesLabel->{"y","x"},
        DisplayFunction->Identity]

-Graphics-

text1 = Graphics[
        Text[FontForm["(x1,y1)",{"Times-Italic",9}],
            {-1,0.5}]];

text2 = Graphics[
        Text[FontForm["(x2,y2)",{"Times-Italic",9}],
            {7,-2Pi}]];

<<Graphics`Arrow`

pl2 = Graphics[Arrow[{.5,-2},{.5,-4}]]

-Graphics-

text3 = Graphics[
        Text[FontForm["F",{"Times-Bold",9}],
            {0.6,-4.5}]];

Show[{pl1,pl2,text1,text2,text3},
     DisplayFunction->$DisplayFunction]

-Graphics-
```

We also assume that the motion is free of damping and thus the total energy of the particle is conserved; i.e., the sum of potential and kinetic energy remains constant.

$$T + U = const. \tag{2.107}$$

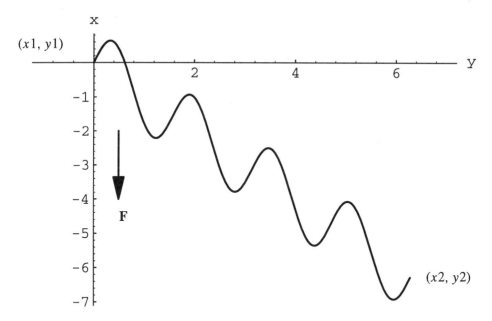

Figure 2.30. Geometry of the Brachistochrone problem.

We choose the origin of the potential energy in such a way that the starting point of the particle is located at the minimum of the potential.

$$U(x = 0) = 0. \tag{2.108}$$

The potential energy thus reads

```
U = -F x = -m g x
```

where g is the acceleration of gravity. The kinetic energy is determined by the velocity of the particle.

```
T = m v^2 / 2
```

The conservation of energy and the fact that the motion starts at rest $(T+U = 0)$ gives us for the velocity **vh**

```
vh = Solve[T+U == 0,v]

{{v -> -(Sqrt[2] Sqrt[g] Sqrt[x])},

 {v -> Sqrt[2] Sqrt[g] Sqrt[x]}}
```

The time needed to move from point p_1 to point p_2 is given by the integral of the path

$$t = \int_{x_1,x_2}^{y_1,y_2} \frac{ds}{v} \tag{2.109}$$

where $ds = (dx^2 + dy^2)^{(1/2)}$ corresponds to the line element in Cartesian coordinates. The functional density ds/v being equal to the differential of time dt is given by

```
dt=Sqrt[(1+D[y[x],x]^2)/x];
```

We have suppressed the constant $1/Sqrt[2g]$ since it does not change the final result of the Brachistochrone problem. We determine the variation of the density by replacing $y[x]$ and its derivatives with the function $y = y + ew$. Differentiating action S with respect to parameter e and then setting $e = 0$ gives us the equation of motion. Simplifying the calculation we define function y to be

```
rule1 = y->Function[x,y[x] + e w[x]]

y -> Function[x, y[x] + e w[x]]
```

The application of **rule1** to the density **dt** gives us

```
dth = dt /.rule1

                          2
        1 + (e w'[x] + y'[x])
Sqrt[---------------------]
               x
```

Differentiating with respect to e and setting $e = 0$ delivers

```
dth = D[dth,e] /. e->0

   w'[x] y'[x]
   -----------------
                 2
        1 + y'[x]
   x Sqrt[----------]
            x
```

The result contains derivatives of test function w. Since the whole expression is part of an integral, we can simplify the integrand by integration of parts. For our calculation we use the fundamental lemma of the calculus of variations. The lemma states that the test function vanishes at the boundaries. These conditions allow the transformation of the integral to a standard representation.

Unfortunately the standard *Mathematica* function **Integrate[]** does not perform an integration by parts. Therefore we have to redefine **Integrate[]** by the following expressions.

```
Unprotect[Integrate]

{Integrate}

Integrate[b_. Derivative[n_][w][x],x]:=
Integrate[w[x] (-1)^n Hold[D[b,{x,n}]],x]
```

First we un-protect **Integrate[]**; then we define the new rules for our calculation and finally, we protect **Integrate[]** again. Note that **Hold[]** suppresses an explicit integration of derivatives.

```
Protect[Integrate]

{Integrate}
```

The integration of the expression **dth** now gives us

```
dth = Integrate[dth,x]

                      y'[x]
   -Integrate[Hold[D[-----------------, {x, 1}]] w[x], x]
                                  2
                        1 + y'[x]
              x Sqrt[----------]
                        x
```

Since the test function is an arbitrary function used to change y, we are only interested in the coefficient of w. The coefficient can be extracted by

```
dth = Coefficient[dth[[2,1]],w[x]]

              y'[x]
  Hold[D[------------------, {x, 1}]]
                          2
                  1 + y'[x]
          x Sqrt[----------]
                     x
```

The term **dth** disappears since the integral given above vanishes according to the calculus of variations. Thus we extract the needed term from **dth** and set it equal to the integration constant $1/\sqrt{2a}$.

```
dth = dth[[1,1]]

          y'[x]
  ------------------
                  2
          1 + y'[x]
  x Sqrt[----------]
             x

dth = dth == 1/Sqrt[2 a]

          y'[x]                    1
  ------------------ == ---------------
                  2       Sqrt[2] Sqrt[a]
          1 + y'[x]
  x Sqrt[----------]
             x
```

Squaring both sides of this equation we get a differential equation which can be solved by integration

```
dth = Thread[dth^2,Equal]

            2
        y'[x]              1
  -------------- == ---
                2      2 a
  x (1 + y'[x] )


dthh = Solve[dth,D[y[x],x]]
```

```
                 Sqrt[x]                          Sqrt[x]
{{y'[x] -> -(-------------)},  {y'[x] -> -------------}}
               Sqrt[2 a - x]                    Sqrt[2 a - x]
```

The integrand is represented by

```
int = D[y[x],x] /. dthh[[2]]

   Sqrt[x]
-------------
Sqrt[2 a - x]
```

The derived expression represents the integrand of the action integral. We simplify
the integrand to a more manageable form in *Mathematica* by substituting

```
subst1 = x->a(1-Cos[theta]);
```

The differential dx is replaced by the new differential *theta* multiplied by a factor.

```
dx = D[x/.subst1,theta]

a Sin[theta]
```

The integrand in the updated variables is given by

```
ints = dx int/.subst1

a Sqrt[a (1 - Cos[theta])] Sin[theta]
-------------------------------------
   Sqrt[2 a - a (1 - Cos[theta])]
```

This expression is simplified by the following chain of functions to

```
ints = Sqrt[ints^2//PowerExpand//Simplify]//PowerExpand

          theta 2
2 a Sin[-----]
            2
```

which can easily be integrated with the result

```
y = Integrate[ints,theta]

a theta - a Sin[theta]
```

We now know that the path between p_1 and p_2 in a parametric form given by the coordinates x and y depends on *theta*. x and y describe the fastest connection between two points in a homogeneous force field. Parameter a contained in the representation above has been adjusted so that the path of the particle passes point p_2. The curve derived is known as a cycloid.

```
curve = {y,-x/.subst1}

{a theta - a Sin[theta], -(a (1 - Cos[theta]))}
```

A parametric representation of the solution for different parameters a is created by the function **ParametricPlot[ ]** and given below

```
k1 = curve /. a->1;
k2 = curve /. a->2;
k3 = curve /. a->1.25;
k4 = curve /. a->1.5;

ParametricPlot[{k1,k2,k3,k4},{theta,0,2Pi}]

-Graphics-
```

The classical problem of the Brachistochrone deals with a variational problem and the necessary steps to calculate its functional derivative. We summarize the main steps of our calculation in order to demonstrate the algorithmic procedure needed to derive equations of motion.

1. Replacement of the dependent function y with its variation $y = y + ew$.

2. Differentiation of the functional density with respect to the arbitrary parameter e and replacement of e with zero after the differentiation.

3. Application of the fundamental lemma to simplify all differentiations of the test function in the integrand.

4. The coefficients of the test function w deliver the equation of motion which determines the track of a particle.

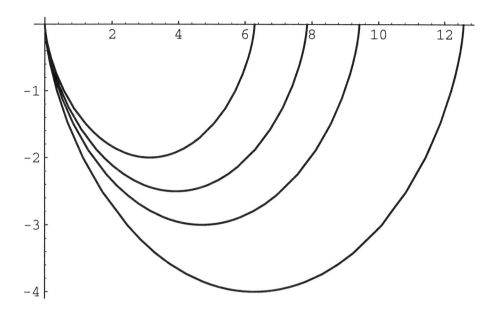

Figure 2.31.   Solutions of the Brachistochrone problem for different values of parameter a.

The equations derived by these steps are called Euler–Lagrange equations and the basis of the calculation is the Lagrangian.

The function **EulerLagrange**[ ] implements the four steps of calculating the Euler–Lagrange equation for a point system. In this context the **EulerLagrange**[ ] function is defined for systems depending only on one independent and one dependent variable. Steps 3) and 4) of the algorithm simulate an integration by parts. This means that the derivatives of the test function w are transformed to the part of the integrand which is independent of w. Thus we can replace the integration by parts by differentiating the integrand.

```
Options[EulerLagrange]={eXpand->False}

{eXpand -> False}

EulerLagrange[density_,depend_,independ_,options___]:=
        Block[{f0,rule,fh,e,w,y,expand},
(* --- check the options --- *)
        {expand} = {eXpand} /. {options} /.
                            Options[EulerLagrange];
(* --- rule for the variation of y --- *)

        f0 = Function[x,y[x] + e w[x]];

(* --- rules for the replacement of derivatives of the
        test function. We use a delayed rule to
```

```
        prevent an evaluation of the expressions.
    --- *)

        rule = b_. Derivative[n_][w][independ]:>
               (-1)^n Hold[D[b,{independ,n}]];

(* --- replacement of y by y + e w ---*)

        fh = density /. depend -> f0 /. {x->independ,
                            y->depend};

(* --- differentation with respect to e --- *)
        fh = D[fh,e] /. e->0 // Expand;
(* --- transform derivatives of w to
        derivatives of y ---*)
        fh = fh /. rule /. w[independ]->1;
(* --- calculate the Euler--Lagrange equations --- *)
        If[expand,
            fh = fh // ReleaseHold,
            fh
        ]
]
```

If we apply the **EulerLagrange[ ]** function to a general Lagrangian depending on higher derivatives of the dependent function, we obtain the general representation of the Euler–Lagrange equations.

As a first example we demonstrate the application of **EulerLagrange[ ]** for a density depending only on first derivatives with respect to time t. To represent such a density in *Mathematica* we define the Lagrangian by

```
l = L[q[t],D[q[t],t]]

L[q[t], q'[t]]
```

The application of the **EulerLagrange[ ]** function yields the expression

```
elgl = EulerLagrange[l,q,t]

        (0,1)
-Hold[D[L     [q[t], q'[t]], {t, 1}]] +
   (1,0)
  L     [q[t], q'[t]]
```

This is the general representation of the Euler–Lagrange equation for a Lagrangian depending only on generalized velocities and coordinates. The resulting representation is equivalent to the general expression known from classical mechanics [2.3]

$$\frac{\partial L}{\partial q} - \frac{d}{dt}\left(\frac{\partial L}{\partial \dot{q}}\right) = 0. \tag{2.110}$$

The **EulerLagrange**[ ] function delivers a representation in an unevaluated form preventing its explicit differentiation with respect to time. Using **ReleaseHold**[ ] expands this expression to a standard representation

```
elgl // ReleaseHold
          (0,2)
 -(q''[t] L      [q[t], q'[t]]) +
    (1,0)
    L      [q[t], q'[t]] -
          (1,1)
    q'[t] L      [q[t], q'[t]]
```

The expansion of this result can also be obtained by using the **EulerLagrange**[ ] function. To activate the expansion property of the **EulerLagrange**[ ] function, we have to specify the option **eXpand** as its fourth argument. Setting **eXpand** to the value **True** we get

```
EulerLagrange[l,q,t,eXpand->True]
          (0,2)
 -(q''[t] L      [q[t], q'[t]]) +
    (1,0)
    L      [q[t], q'[t]] -
          (1,1)
    q'[t] L      [q[t], q'[t]]
```

The default value of the option **eXpand** is set to **False**; i.e., all differentiations are held.

Availing of the **EulerLagrange**[ ] function and knowing the Lagrangian, we are able to derive any equation of motion. As an example we consider a Lagrangian depending on derivatives up to second order:

```
l1 = L[q[t],D[q[t],t],D[q[t],{t,2}]]

L[q[t], q'[t], q''[t]]
```

Applying **EulerLagrange**[ ], we get

```
EulerLagrange[ll,q,t]

        (0,0,1)
 Hold[D[L        [q[t], q'[t], q''[t]],
                           (0,1,0)
     {t, 2}]] - Hold[D[L        [q[t], q'[t],
      q''[t]], {t, 1}]] +
   (1,0,0)
    L       [q[t], q'[t], q''[t]]
```

which is an ordinary differential equation of fourth order.

2.8.1.3 Harmonic oscillator
The Lagrangian of the harmonic oscillator is given by the difference between the kinetic and the potential energy

```
L = T - U
```

Expressing the kinetic energy in generalized coordinates, T is represented by

```
T = m D[q[t],t]^2/2

        2
 m q'[t]
 --------
    2
```

The potential energy U is

```
U = k q[t]^2/2

        2
 k q[t]
 -------
    2
```

Consequently, the related Lagrangian is

```
  l = T-U

       2           2
  -(k q[t] )   m q'[t]
  ---------- + --------
      2           2
```

The Euler–Lagrange equations of the harmonic oscillator can be derived using

```
  harm = EulerLagrange[l,q,t]//ReleaseHold

  -(k q[t]) - m q''[t]
```

Scaling the time by the factor $(k/m)^{1/2}$, we can reduce the equation of motion to a form free of any parameters.

```
  harm = harm == 0 /.{k->1,m->1}

  -q[t] - q''[t] == 0
```

Solving the equation using **DSolve**[ ] and the initial conditions $q(0) = 1$ and $q'(0) = -1$, we get the solution

```
  sol = DSolve[harm,q[t],t]

  {{q[t] -> Cos[t] - Sin[t]}}
```

We use **Plot**[ ] to represent the solution, shown in figure 2.32.

```
  Plot[Evaluate[q[t] /. sol[[1]]],{t,0,2Pi},AxesLabel->{"t","q"},
          Frame->True,
          PlotLabel->"Harmonic oscillator"]

  -Graphics-
```

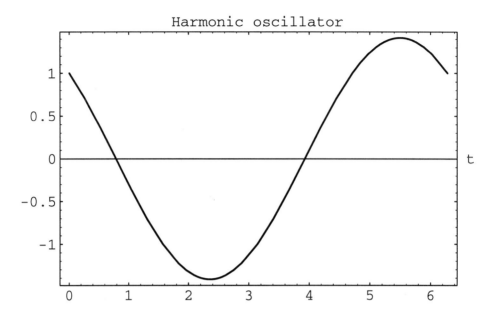

Figure 2.32.   Solutions of the harmonic oscillator.

2.8.1.4 Anharmonic oscillator   Another example of the Lagrange formalism is
the derivation of the equations of motion for an anharmonic oscillator. In an earlier
section, we discussed the various types of motion of the pendulum. As with the
harmonic oscillator, the Lagrangian of the anharmonic oscillator is determined by
the difference between kinetic and potential energy.

The anharmonic oscillator or pendulum consists of a bob with mass m. m is
located at the end of a massless rod of length a. This rod may but need not rotate
around a supporting point in the gravitational field of the earth. The generalized
coordinate of the pendulum is the angle of the rod with respect to the vertical axis
of the coordinate system (see figure 2.33).

```
<<Graphics'Arrow'

lines = {Line[{{-2,0},{2,0}}],Line[{{0,0},{0,-2}}],
        Line[{{0,0},{-3,-5}}]};

arow = Arrow[{-3,-5},{-1.2,-6.1}];

circ = Circle[{-3,-5},1/2];

t1 = Text[FontForm["F",{"Times-Bold",12}],{-1.2,-6.8}];

t2 = Text[FontForm["m",{"Times-Bold",12}],{-3.5,-6}];

t3 = Text[FontForm["a",{"Times-Bold",12}],{-2.2,-2.5}];
```

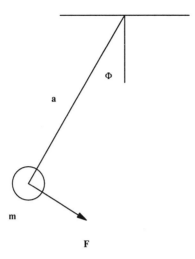

Figure 2.33.　The pendulum.

```
t4 = Text[FontForm["F",{"Symbol",12}],{-0.5,-1.8}];

Show[Graphics[{lines,arow,circ,t1,t2,t3,t4}],AspectRatio->Automatic]

-Graphics-
```

We can determine the potential energy from the geometry of the pendulum by

```
U = m g a ( 1 - Cos[Phi[t]])

a g m (1 - Cos[Phi[t]])
```

The kinetic energy is given by

```
T = m a^2 D[Phi[t],t]^2/2

 2      2
a  m Phi'[t]
-------------
      2
```

Using U and T, we define the Lagrangian by

```
Lanharm = T - U

                                    2       2
                                 a  m Phi'[t]
 -(a g m (1 - Cos[Phi[t]])) + -------------
                                     2
```

Again, the equation of motion is derived by using the **EulerLagrange**[ ] function. The fourth argument of **EulerLagrange**[ ] is set to True to get a full expansion of the equation.

```
anharmgl = EulerLagrange[Lanharm,Phi,t,eXpand->True]

                         2
 -(a g m Sin[Phi[t]]) - a  m Phi''[t]
```

We get the standard form of this equation if we divide the equation by $-a^2 m$:

```
anharmgl = anharmgl/(- a^2 m) // Expand

g Sin[Phi[t]]
------------- + Phi''[t]
      a
```

We solve this nonlinear second order differential equation by using **DSolve**[ ]. The function **DSolve**[ ] defined in the kernel of *Mathematica* is not capable of solving this equation. However, if we load the standard packages **Calculus'DSolve'** and **Calculus'EllipticIntegrate'**, we obtain an implicit solution of the equation.

```
<<Calculus'DSolve'

<<Calculus'EllipticIntegrate'

DSolve[anharmgl==0,Phi[t],t]

                                  -C[1] + Cos[Phi[t]]
{Solve[-((Sqrt[2] Sqrt[a] Sqrt[-------------------]
                                       1 - C[1]

              Phi[t]    2
    EllipticF[------, --------]) /
                2     1 - C[1]

    (Sqrt[g] Sqrt[-C[1] + Cos[Phi[t]]])) == t + C[2], Phi[t]],
```

```
                           -C[1] + Cos[Phi[t]]
  Solve[(Sqrt[2] Sqrt[a] Sqrt[-------------------]
                             1 - C[1]

              Phi[t]      2
    EllipticF[------, --------]) /
                2      1 - C[1]

  (Sqrt[g] Sqrt[-C[1] + Cos[Phi[t]]]) == t + C[2], Phi[t]]}
```

The solution shows that we need the inverse function of the elliptic integrals. *Mathematica* recognizes this type of function as the Jacobi elliptic function. The disadvantage of *Mathematica* 2.2 is that it is not capable of linking the elliptic integrals with the Jacobi elliptic functions. We would have to establish this link by hand in order to obtain an explicit solution. However, our intention is not to derive the explicit solution of the pendulum but to give a dynamic formulation of the equation of motion by means of the Euler–Lagrange formalism.

The above two examples demonstrate the derivation of the equations of motion for systems with only one generalized coordinate (q) depending on one independent variable (t). In the next section, we generalize the **EulerLagrange[]** function for systems containing more than one dependent variable.

2.8.2 Lagrangian depending on a set of variables

In the previous section, we derived the Euler–Lagrange equations by means of a variation in one dependent variable $q(t)$. However, in physics, we do not usually encounter situations in which equations are represented by only one dependent variable but rather by a set of dependent variables q. For example, a Lagrangian depending on several dependent variables is

$$L = L(q_1, q_2, \ldots, q'_1, q'_2, \ldots). \tag{2.111}$$

If we again use the calculus of variations to derive the equations of motion, we have to take into account the variation of each variable q_i by

$$q_i(x, e) = q_i(x) + ew_i(x) \qquad i = 1, 2, \ldots \tag{2.112}$$

The remaining steps in the calculation of the Euler–Lagrange equations are the same as those for one dependent variable. We only need to extend our **EulerLagrange[]** function to individual variations. The following listing gives the extended version of this function. A loop is included in this list which scans all the dependent variables.

```
EulerLagrange[density_,depend_List,independ_,
              options___]:=
        Block[{f0,fh,e,w,y,expand,euler={},
              wtable},
(* --- check the options ---*)
        {expand} = {eXpand} /. {options} /.
                          Options[EulerLagrange];
(* --- table of test functions w --- *)
        wtable = Table[w[i],{i,1,Length[depend]}];
(* --- replacement rule for y --- *)

        f0 = Function[x,y[x] + e w[x]];

(* --- rules to replace the derivatives in
       the test functions by derivatives
       of the rest of the integrand    --- *)

        rules[i_] := b_. Derivative[n_][wtable[[i]]][independ]:>
              (-1)^n Hold[D[b,{independ,n}]]];
(* --- loop over dependent variables --- *)
Do[
(* --- replace y by the variation ---*)

        fh = density /. depend[[j]] -> f0 /.
                              {x->independ,
                               y->depend[[j]],
                               w->wtable[[j]]};

(* --- differentiation with respect to e --- *)
        fh = D[fh,e] /. e->0 // Expand;
(* --- use the fundamental lemma of variation ---*)
        fh = fh /. rules[j] /.
                      wtable[[j]][independ]->1;
        AppendTo[euler,fh],
{j,1,Length[depend]}];
(* --- calculate the Euler--Lagrange equations --- *)
        If[expand,
           euler = euler // ReleaseHold,
           euler
          ]
]
```

To check the **EulerLagrange[]** function, let us look at a Lagrangian depending on two generalized coordinates. The Lagrangian in q_1 and q_2 is given by the expression

```
lag = L[q1[t],q2[t],D[q1[t],t],D[q2[t],t]]

L[q1[t], q2[t], q1'[t], q2'[t]]
```

The corresponding equations of motion are derived by

```
EulerLagrange[lag,{q1,q2},t]

            (0,0,1,0)
 {-Hold[D[L           [q1[t], q2[t], q1'[t],
       q2'[t]], {t, 1}]] +
    (1,0,0,0)
    L            [q1[t], q2[t], q1'[t], q2'[t]],
            (0,0,0,1)
   -Hold[D[L           [q1[t], q2[t], q1'[t],
       q2'[t]], {t, 1}]] +
    (0,1,0,0)
    L           [q1[t], q2[t], q1'[t], q2'[t]]}
```

The second argument of **EulerLagrange[ ]** now requires a list for the dependent variables. The result is a system of equations for q_1 and q_2. If we expand these equations, we get a coupled second order system of ordinary differential equations. Comparing the result with the standard mathematical notation given by

$$\frac{\partial L}{\partial q_1} - \frac{d}{dt}\left(\frac{\partial L}{\partial \dot{q}_1}\right) = 0 \qquad (2.113)$$

$$\frac{\partial L}{\partial q_2} - \frac{d}{dt}\left(\frac{\partial L}{\partial \dot{q}_2}\right) = 0 \qquad (2.114)$$

we observe that the new version of **EulerLagrange[ ]** delivers the left hand sides of the equations. These equations are equivalent to a coupled second order system of ordinary differential equations. The expanded form is given here:

```
EulerLagrange[lag,{q1,q2},t,eXpand->True]

              (0,0,1,1)
 {-(q2''[t] L           [q1[t], q2[t], q1'[t],
       q2'[t]]) - q1''[t]
    (0,0,2,0)
    L           [q1[t], q2[t], q1'[t], q2'[t]] -
            (0,1,1,0)
   q2'[t] L           [q1[t], q2[t], q1'[t],
                  (1,0,0,0)
     q2'[t]] + L           [q1[t], q2[t], q1'[t],
       q2'[t]] - q1'[t]
     (1,0,1,0)
    L           [q1[t], q2[t], q1'[t], q2'[t]],
                (0,0,0,2)
   -(q2''[t] L           [q1[t], q2[t], q1'[t],
       q2'[t]]) - q1''[t]
    (0,0,1,1)
    L           [q1[t], q2[t], q1'[t], q2'[t]] +
    (0,1,0,0)
```

```
    L           [q1[t], q2[t], q1'[t], q2'[t]] -
            (0,1,0,1)
  q2'[t] L           [q1[t], q2[t], q1'[t],
    q2'[t]] - q1'[t]
    (1,0,0,1)
    L           [q1[t], q2[t], q1'[t], q2'[t]]}
```

This set of equations is the correct representation of the equations of motion for two generalized coordinates. To see how these equations simplify, we discuss the specific problem of two coupled oscillators.

2.8.2.1 Coupled oscillators   To demonstrate the derivation of systems of equations, let us consider two harmonic oscillators which are coupled by their generalized coordinates, q_1 and q_2. Apart from the harmonic terms, the potential energy contains a coupling consisting of a product of generalized coordinates. The strength of the coupling is determined by the coupling constant k. The Lagrangian of the system is given by the difference between the kinetic and the potential energy. The potential energy is

```
  U = k1 q1[t]^2/2 + k2 q2[t]^2/2 + k q1[t] q2[t]

          2                         2
 k1 q1[t]                  k2 q2[t]
 --------- + k q1[t] q2[t] + ---------
     2                          2
```

where k_1 and k_2 are the spring constants and k represents the coupling strength. The kinetic energy is given by

```
  T = m1 D[q1[t],t]^2/2 + m2 D[q2[t],t]^2/2

          2             2
 m1 q1'[t]     m2 q2'[t]
 ---------- + ----------
     2             2
```

Using T and U we can write down the Lagrangian

```
lagF = T - U

           2                           2
  -(k1 q1[t] )                    k2 q2[t]
  ------------ - k q1[t] q2[t] - --------- +
       2                             2
          2               2
   m1 q1'[t]       m2 q2'[t]
   ----------- +  -----------
       2               2
```

The equations of motion result from using the **EulerLagrange[ ]** function in the form shown here:

```
glsys = EulerLagrange[lagF,{q1,q2},t,eXpand->True]

{-(k1 q1[t]) - k q2[t] - m1 q1''[t],
   -(k q1[t]) - k2 q2[t] - m2 q2''[t]}
```

The resulting system of equations is a linear coupled system of second order containing the spring constants k_1, k_2 and k, and the masses m_1 and m_2 as parameters.

2.8.2.2 The double pendulum   Another non-trivial example of formulating equations of motion using the Euler–Lagrange formalism is given by the double pendulum. The double pendulum is a pendulum consisting of two masses and two rods which are connected by mass m_1 (see figure 2.34).

```
(* create a graphical representation of the double pendulum *)
<<Graphics'Arrow'

(* represent the pendulas by lines *)
lines = {Line[{{-2,0},{0,0}}],Line[{{0,0},{-1,-3}}],
         Line[{{-1,-3},{-2,-7}}],Line[{{0,0},{0,-2}}],
         Line[{{-1,-3},{-1,-5}}],Line[{{0,0},{0,1/2}}]};

(* represent the coordinate axes *)
arrows = {Arrow[{0,0},{2,0}],Arrow[{0,0},{0,-8}]};

(* create two disks for the masse *)
circles = {{RGBColor[1,0,0],Disk[{-1,-3},1/6]},
           {RGBColor[0,1,0],Disk[{-2,-7},1/8]}};

(* annotate the geometrical properties *)
texts = {Text[FontForm["F1",{"Symbol",10}],{-0.3,-2}],
         Text[FontForm["F2",{"Symbol",10}],{-1.2,-5.3}],
         Text[FontForm["m1",{"Helvetica",8}],{-0.5,-3}],
```

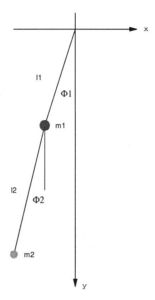

Figure 2.34.   The double pendulum.

```
        Text[FontForm["m2",{"Helvetica",8}],{-1.5,-7}],
        Text[FontForm["l1",{"Helvetica",8}],{-1.2,-1.5}],
        Text[FontForm["l2",{"Helvetica",8}],{-2,-5}],
        Text[FontForm["x",{"Italic-Bold",8}],{2.3,0}],
        Text[FontForm["y",{"Italic-Bold",8}],{0.3,-8}]};

(* display the graphics elements *)
Show[Graphics[{lines,arrows,circles,texts}],
     AspectRatio->Automatic]

-Graphics-
```

The generalized coordinates describing the motion of the coupled system of pendula are Φ_1 and Φ_2 (see figure 2.34). We assume that the motion of the two rods is restricted to the (x, y) plane. The Lagrangian of the two pendula is easily derived by the difference between kinetic and potential energy in Cartesian coordinates. The two terms are given by

```
Clear[m1,m2,l1,l2]

T = m1 ( D[x1[t],t]^2 + D[y1[t],t]^2)/2 +
    m2 ( D[x2[t],t]^2 + D[y2[t],t]^2)/2
```

```
       2            2              2            2
 m1 (x1'[t]   + y1'[t] )    m2 (x2'[t]   + y2'[t] )
 ----------------------- +  -----------------------
            2                         2
```

where m_1 and m_2 are the masses of the kinetic energy. The potential energy is represented by

```
 U = -(m1 g y1[t] + m2 g y2[t])

 -(g m1 y1[t]) - g m2 y2[t]
```

where g is the acceleration of gravity. Considering the geometric relations between the Cartesian coordinates and the generalized coordinates given in figure 2.34, we can now introduce a transformation rule to act between the two representations. Transforming the Cartesian to generalized coordinates, we define

```
 coordrule = {x1->Function[t,l1 Sin[Phi1[t]]],
              x2->Function[t,l1 Sin[Phi1[t]] +
                             l2 Sin[Phi2[t]]],
              y1->Function[t,l1 Cos[Phi1[t]]],
              y2->Function[t,l1 Cos[Phi1[t]] +
                             l2 Cos[Phi2[t]]]};
```

Using this rule in the definition of the Lagrangian, we get a representation in the generalized coordinates Φ_1 and Φ_2. The Lagrangian also contains some parameters such as rod lengths l_1 and l_2, masses m_1 and m_2, and acceleration of gravity g

```
 lagrdouble = T - U /. coordrule //Expand

 g l1 m1 Cos[Phi1[t]] + g l1 m2 Cos[Phi1[t]] +
   g l2 m2 Cos[Phi2[t]] +
     2            2           2
   l1   m1 Cos[Phi1[t]]   Phi1'[t]
   ----------------------------- +
               2
     2            2           2
   l1   m2 Cos[Phi1[t]]   Phi1'[t]
   ----------------------------- +
               2
     2            2           2
   l1   m1 Sin[Phi1[t]]   Phi1'[t]
   ----------------------------- +
               2
```

```
     2                2         2
  l1  m2 Sin[Phi1[t]]   Phi1'[t]
  ------------------------------ +
               2
  l1 l2 m2 Cos[Phi1[t]] Cos[Phi2[t]] Phi1'[t]
   Phi2'[t] + l1 l2 m2 Sin[Phi1[t]]
   Sin[Phi2[t]] Phi1'[t] Phi2'[t] +
     2                2         2
  l2  m2 Cos[Phi2[t]]   Phi2'[t]
  ------------------------------ +
               2
     2                2         2
  l2  m2 Sin[Phi2[t]]   Phi2'[t]
  ------------------------------
               2
```

The equations of motion result from applying **EulerLagrange[]** and using **Simplify[]**

```
  gldouble = EulerLagrange[lagrdouble,{Phi1,Phi2},t,
                           eXpand->True]//Simplify

  {-(l1 (g m1 Sin[Phi1[t]] + g m2 Sin[Phi1[t]] +
                                                 2
          l2 m2 Sin[Phi1[t] - Phi2[t]] Phi2'[t]   +
          l1 m1 Phi1''[t] + l1 m2 Phi1''[t] +
          l2 m2 Cos[Phi1[t] - Phi2[t]] Phi2''[t]))\\
     , -(l2 m2 (g Sin[Phi2[t]] -
                                              2
          l1 Sin[Phi1[t] - Phi2[t]] Phi1'[t]   +
          l1 Cos[Phi1[t] - Phi2[t]] Phi1''[t] +
          l2 Phi2''[t]))}
```

The resulting equations of motion are highly nonlinear and coupled. The second order system is too complicated to be solved in a straightforward way.

By using the **EulerLagrange[]** function, we are able to derive all equations of motion for any Lagrangian, provided the Lagrangian depends either on a single or on a set of generalized coordinates depending on one independent variable. These sets of coordinates are important for formulating equations of motion in classical mechanics.

In the following section, we deal with the continuous part of classical mechanics such as continuum mechanics or elasticity theory. In order to treat such theories, the **EulerLagrange[]** function needs to be generalized with respect to the number of independent variables. Generalizing **EulerLagrange[]** opens the door to applications in classical field theory as well.

2.8.3 Lagrange formulation of field equations

We consider the case of several independent and dependent variables. Our intention is to give a dynamic formulation for field equations within the context of the Euler–Lagrange theory. To use the **EulerLagrange**[ ] function in the same manner as before, we need to alter some steps in its function body. The difference in the definitions of the **EulerLagrange**[ ] function is required because we have to consider a set of independent variables. This means that we need to allow the third argument of the **EulerLagrange**[ ] function to be a list. Further, we need to consider the derivatives with respect to the independent variables as partial derivatives. Consequently, we have to consider all possibilities of derivatives occurring under the extended set of independent variables. The changes are listed below. Note that the option **eXpand** has been changed in the **EulerLagrange**[ ] function to mean its opposite because an unexpanded form of the equations is hard to read in the case of field equations. Consequently, the default value in the field formulation will give a completely expanded representation of the equations.

```
EulerLagrange[density_,depend_List,
             independ_List,options___]:=
        Block[{f0,fh,e,w,y,x$m,expand,
                euler={},wtable},
(* --- check the option ---*)
        {expand} = {eXpand} /. {options} /.
                     Options[EulerLagrange];
(* --- table of test functions w --- *)
        wtable = Table[w[i],{i,1,Length[depend]}];
(* --- rule for the variation of y --- *)

        f0 = Function[x$m,y + e w];

(* --- replacement rules for the testfunction w --- *)

ruleg[i_] := b_. Apply[Derivative[n___][wtable[[i]]],
                   independ]:>
      (-1)^Apply[Plus,{n}] Hold[ D[b,
            Delete[Thread[{independ,{n}}],0] ] ];
(* --- loop over all dependent variables --- *)
Do[
(* --- replace y by the variation ---*)

        fh = density /. depend[[j]] -> f0 /.
                   {x$m->independ,
                    y->Apply[depend[[j]],independ],
                    w->Apply[wtable[[j]],independ]};

(* --- differentiation with respect to e --- *)
        fh = D[fh,e] /. e->0 // Expand;
(* --- fundamental lemma of variation ---*)
        fh = fh /. ruleg[j] /.
                   Apply[wtable[[j]],independ]->1;
        AppendTo[euler,fh],
```

```
{j,1,Length[depend]}];
(* --- Calculate the Euler--Lagrange equations --- *)
        If[Not[expand],
            euler = euler //ReleaseHold,
            euler
        ]
]
```

We demonstrate the application of the modified **EulerLagrange**[ ] function by several examples. First we consider a general Lagrangian depending on one field variable q_1, which depends on two independent variables x and t. Assume that the Lagrangian density shows a dependence on the field q_1 and on its first time derivative.

```
la1 = L[q1[x,t],D[q1[x,t],t]]

              (0,1)
L[q1[x, t], q1      [x, t]]
```

The form of the Euler–Lagrange equations results in

```
EulerLagrange[la1,{q1},{x,t},eXpand->True]

          (0,1)                (0,1)
{-Hold[D[L     [q1[x, t], q1      [x, t]],
     Delete[Thread[{{x, t}, {0, 1}}], 0]]] +
   (1,0)                (0,1)
  L      [q1[x, t], q1      [x, t]]}
```

Its completely expanded form is given by

```
EulerLagrange[la1,{q1},{x,t}]

     (0,2)                (0,1)
{-(L      [q1[x, t], q1      [x, t]]
      (0,2)
    q1      [x, t]) +
   (1,0)                (0,1)
  L      [q1[x, t], q1      [x, t]] -
    (0,1)          (1,1)                (0,1)
  q1      [x, t] L      [q1[x, t], q1      [x, t]]
  }
```

The resulting equation is a partial differential equation of second order with respect to time. Its dependence on the first order time derivative will result in second order equations. To demonstrate this type of behavior let us again consider a Lagrangian depending on squares of the gradients with respect to time and space. We define

```
la2 = D[u[x,t],t]^2/2 + D[u[x,t],x]^2/2

 (0,1)      2   (1,0)      2
u     [x, t]   u     [x, t]
------------- + -------------
      2               2
```

The **EulerLagrange[ ]** function gives us the equation of motion

```
EulerLagrange[la2,{u},{x,t}]

   (0,2)         (2,0)
{-u    [x, t] - u    [x, t]}
```

which is simply a scalar wave equation in u with constant propagation velocity $c = 1$. This kind of wave equation is well known in physics. Its three dimensional form occurs in electrodynamics for the electric or magnetic field. To generalize the above equation for $3 + 1$ dimensions let us consider the Lagrangian

```
la3 = (D[u[x1,x2,x3,t],x1]^2 +
       D[u[x1,x2,x3,t],x2]^2 +
       D[u[x1,x2,x3,t],x3]^2 +
       D[u[x1,x2,x3,t],t]^2)/2

  (0,0,0,1)             2
(u         [x1, x2, x3, t]  +
  (0,0,1,0)             2
 u         [x1, x2, x3, t]  +
  (0,1,0,0)             2
 u         [x1, x2, x3, t]  +
  (1,0,0,0)             2
 u         [x1, x2, x3, t] ) / 2
```

which provides the equation of motion

```
EulerLagrange[la3,{u},{x1,x2,x3,t}]

   (0,0,0,2)
```

```
{-u          [x1, x2, x3, t] -
   (0,0,2,0)
  u          [x1, x2, x3, t] -
   (0,2,0,0)
  u          [x1, x2, x3, t] -
   (2,0,0,0)
  u          [x1, x2, x3, t]}
```

Another case of interest is the occurrence of more than one field in the Lagrangian.
For such a case we get a system of equations. Let us thus introduce a second field
variable in the Lagrangian, **la3**, by replacing u with v.

```
la4 = la3 + (la3 /.u->v)

   (0,0,0,1)                2
(u          [x1, x2, x3, t]  +
   (0,0,1,0)                2
  u          [x1, x2, x3, t]  +
   (0,1,0,0)                2
  u          [x1, x2, x3, t]  +
   (1,0,0,0)                2
  u          [x1, x2, x3, t] ) / 2 +
   (0,0,0,1)                2
(v          [x1, x2, x3, t]  +
   (0,0,1,0)                2
  v          [x1, x2, x3, t]  +
   (0,1,0,0)                2
  v          [x1, x2, x3, t]  +
   (1,0,0,0)                2
  v          [x1, x2, x3, t] ) / 2
```

The resulting system of equations reads

```
EulerLagrange[la4,{u,v},{x1,x2,x3,t}]

   (0,0,0,2)
{-u          [x1, x2, x3, t] -
   (0,0,2,0)
  u          [x1, x2, x3, t] -
   (0,2,0,0)
  u          [x1, x2, x3, t] -
   (2,0,0,0)
  u          [x1, x2, x3, t],
   (0,0,0,2)
 -v          [x1, x2, x3, t] -
   (0,0,2,0)
  v          [x1, x2, x3, t] -
   (0,2,0,0)
  v          [x1, x2, x3, t] -
```

```
     (2,0,0,0)
    v          [x1, x2, x3, t]}
```

representing a decoupled system of wave equations in $3+1$ dimensions. Since we have used the Lagrangian free of any coupling between the fields u and v, we can only derive a decoupled system of equations. If we consider the same Lagrangian and add a coupling term by multiplying the fields by each other, we get

```
la5 = la4 + u[x1,x2,x3,t] v[x1,x2,x3,t]

u[x1, x2, x3, t] v[x1, x2, x3, t] +
    (0,0,0,1)                2
   (u         [x1, x2, x3, t]  +
     (0,0,1,0)                2
    u           [x1, x2, x3, t]  +
     (0,1,0,0)                2
    u           [x1, x2, x3, t]  +
     (1,0,0,0)                2
    u           [x1, x2, x3, t] ) / 2 +
    (0,0,0,1)                2
   (v         [x1, x2, x3, t]  +
     (0,0,1,0)                2
    v           [x1, x2, x3, t]  +
     (0,1,0,0)                2
    v           [x1, x2, x3, t]  +
     (1,0,0,0)                2
    v           [x1, x2, x3, t] ) / 2
```

which results in a coupled system of wave equations

```
EulerLagrange[la5,{u,v},{x1,x2,x3,t}]

                          (0,0,0,2)
{v[x1, x2, x3, t] - u           [x1, x2, x3, t] -
    (0,0,2,0)
   u           [x1, x2, x3, t] -
    (0,2,0,0)
   u           [x1, x2, x3, t] -
    (2,0,0,0)
   u           [x1, x2, x3, t],
   u[x1, x2, x3, t] -
    (0,0,0,2)
   v           [x1, x2, x3, t] -
    (0,0,2,0)
   v           [x1, x2, x3, t] -
    (0,2,0,0)
   v           [x1, x2, x3, t] -
    (2,0,0,0)
   v           [x1, x2, x3, t]}
```

The simplicity of these examples shows that it is fairly easy to derive complicated equations of motion by using the **EulerLagrange**[ ] function. The real challenge in formulating the equations of motion is the derivation of the Lagrangian. Once we know the Lagrangian, the equations of motion are derived by applying the **EulerLagrange**[ ] function no matter how simple or complicated the Lagrangian may be. If we understand the Lagrangian in terms of kinetic and potential energy, we are able to give a dynamic formulation within the framework of Lagrange dynamics. Another point is that we can easily change the character of the model by changing the Lagrangian. To demonstrate the potential of our formulation we consider some popular nonlinear field models.

2.8.3.1 Nonlinear field equations   To derive one of the many nonlinear field equations let us consider the Lagrangian in the form

```
   nonlag = D[u[x,t],t]^2/2 +
            D[u[x,t],x]^3 u[x,t]^2

    (0,1)        2
   u      [x, t]                2  (1,0)        3
   -------------  + u[x, t]   u       [x, t]
        2
```

which contains the square of the first time derivative and couples the field variable u with the cube of the gradient in $1 + 1$ dimensions. The resulting field equation is of second order in time and space and connects the field with the gradients.

```
   EulerLagrange[nonlag,{u},{x,t}]

      (0,2)                  (1,0)        3
   {-u      [x, t] - 4 u[x, t] u       [x, t]  -
                 2  (1,0)          (2,0)
       6 u[x, t]  u       [x, t] u        [x, t]}
```

If we change the Lagrangian so that its fields are replaced by higher order derivatives with respect to x, we get a change in the resulting field equation

```
   nonlag1 = D[u[x,t],t]^2/2 +
             D[u[x,t],x]^3 D[u[x,t],{x,2}]^2

    (0,1)         2
   u      [x, t]        (1,0)        3  (2,0)         2
   -------------  + u        [x, t]  u       [x, t]
         2
```

```
EulerLagrange[nonlag1,{u},{x,t}]

     (0,2)              (1,0)         (2,0)        3
{-u     [x, t] + 6 u      [x, t] u      [x, t]  +
      (1,0)         2 (2,0)          (3,0)
   12 u      [x, t] u      [x, t] u      [x, t] +
      (1,0)         3 (4,0)
    2 u      [x, t] u      [x, t]}
```

The resulting equation is of fourth order in x. Another example for a system of equations follows from the Lagrangian

```
slag = v[x,t] D[u[x,t],t] +
       D[u[x,t],x] D[v[x,t],x]+
       V[x] u[x,t] v[x,t] +
       u[x,t]^2 v[x,t]^2

        2          2
u[x, t]   v[x, t]  + u[x, t] v[x, t] V[x] +
           (0,1)
   v[x, t] u      [x, t] +
   (1,0)          (1,0)
   u      [x, t] v      [x, t]
```

The equations of motion are

```
EulerLagrange[slag,{u,v},{x,t}]

                   2
{2 u[x, t] v[x, t]  + v[x, t] V[x] -
    (0,1)           (2,0)
   v      [x, t] - v      [x, t],
                 2
   2 u[x, t]  v[x, t] + u[x, t] V[x] +
     (0,1)           (2,0)
   u      [x, t] - u      [x, t]}
```

This system of equations includes a cubic nonlinearity. It is closely related to the cubic nonlinear Schrödinger equation (NLS). The NLS equation is used to describe the interaction of particles via a hard core interaction in quantum field theories or to examine the electrical field amplitude in nonlinear media in optics.

Another well-known example is the Sine–Gordon equation (SG). The SG equation is an equation long known in solid state physics where it is used to study the dislocations in solids. Today the SG equation is one of the nonlinear partial differential equations which serves to examine the properties of solitary waves. We give here the dynamic formulation of the SG equation in Lagrangian form:

```
lsin = (D[u[x,t],t]^2 + D[u[x,t],x]^2)/2 -
        Cos[u[x,t]]

                  (0,1)     2    (1,0)     2
                 u    [x, t]  + u     [x, t]
   -Cos[u[x, t]] + ----------------------------
                              2
```

The related equation of motion follows from

```
EulerLagrange[lsin,{u},{x,t}]

              (0,2)           (2,0)
   {Sin[u[x, t]] - u    [x, t] - u    [x, t]}
```

Another well known nonlinear partial differential equation coming originally from hydrodynamics is the Korteweg deVries (KdV) equation. The KdV equation describes shallow water waves in channels. It is also one of the general nonlinear equations which follows from approximations of several nonlinear field models. The KdV equation is used to describe models in solid state physics, plasma physics, hydrodynamics, etc. It is usually assumed that the KdV equation is not derivable from a Lagrangian. We demonstrate here that a system of equations containing the KdV equation in a hidden form is derivable from a Lagrangian. To give a dynamic formulation of the KdV equation we start out from the Lagrangian for two field variables u and v

```
lkdv = (D[u[x,t],t] D[u[x,t],x])/2 +
        D[u[x,t],x]^3/6 +
        D[u[x,t],x] D[v[x,t],x] +
        v[x,t]^2/2

        2    (0,1)          (1,0)
   v[x, t]  u    [x, t] u     [x, t]
   -------- + ----------------------- +
      2                  2
    (1,0)     3
   u    [x, t]   (1,0)          (1,0)
   ------------- + u    [x, t] v    [x, t]
        6
```

The system of coupled equations results from

```
glkdv = EulerLagrange[lkdv,{u,v},{x,t}]
```

```
      (1,1)          (1,0)          (2,0)
{-u      [x, t] - u      [x, t] u      [x, t] -
    (2,0)                      (2,0)
    v      [x, t], v[x, t] - u      [x, t]}
```

If we eliminate the variable v and replace the spatial derivatives by a function w we get

```
sub1 = v->Function[{x,t},D[u[x,t],{x,2}]];

glkdv = glkdv /.sub1

      (1,1)          (1,0)          (2,0)
{-u      [x, t] - u      [x, t] u      [x, t] -
    (4,0)
    u      [x, t], 0}

sub2 = b_. Derivative[n_,m_][u][x,t]:>
        b Derivative[n-1,m][w][x,t]

          (n_,m_)                (n - 1,m)
(b_.) u       [x, t] :> b w          [x, t]

glkdv = glkdv //.sub2

      (0,1)                (1,0)
{-w      [x, t] - w[x, t] w      [x, t] -
    (3,0)
    w      [x, t], 0}
```

which is the KdV equation in its normalized form.

The examples given in this section have demonstrated that the Euler–Lagrange formulation is quite simple, provided we have a function which calculates the Euler–Lagrange equations. With this function and the Lagrangian of the physical problem, we are able to derive any equation of motion.

2.9 Exercises

1. Examine the fundamental properties of the Poisson bracket

$$\{q_i, p_j\} = \begin{cases} 1 & \text{for} \quad i = j \\ 0 & \text{for} \quad i \neq j \end{cases}$$

$$\{q_i, q_j\} = \{p_i, p_j\} = 0.$$

Use the **Poisson** package.

2. Prove that the Jacobi identity

$$\{A,\{B,C\}\} + \{B,\{C,A\}\} + \{C,\{A,B\}\} = 0$$

   is satisfied for functions A, B, and C depending on the canonical variables $q_1, q_2, \ldots, q_n, p_1, p_2, \ldots, p_n$.

   Hint: An arbitrary function in *Mathematica* is defined by $\mathbf{A} = \mathbf{A[q1,q2,-p1,p2]}$.

3. Use the **Poisson** package to derive the equations of motion for the Hamiltonian

$$H = \frac{1}{2}(p_1^2 + p_2^2) + \frac{1}{2}(q_1^2 + \lambda^2 q_2^2) + \frac{1}{(q_1^2 + q_2^2)^{1/2}}.$$

4. Prove that the following expression

$$I = \left(p_1 p_2 q_2 - q_1 p_2^2 + \frac{1}{4} q_2^2 q_1 - \frac{q_1}{(q_1^2 + q_2^2)^{1/2}} \right) \qquad (2.115)$$

   with $\lambda = \pm\frac{1}{2}$ is a second integral of motion belonging to the Hamiltonian as given in exercise 3.

   Hint: Integrals of motion are defined by $\dot{I} = \{I, H\} = 0$.

5. Reconsider the problem of two coupled harmonic oscillators discussed in Section 2.1 if all three springs have the same spring constant. Determine the characteristic frequencies and compare the magnitudes with the frequencies calculated in the text.

6. Consider two identical, coupled harmonic oscillators (as shown in figure 2.1). Let the oscillator with mass m_2 be damped. The damping parameter is γ. A force $F_0 \cos \omega t$ is applied to m_1. Write down the pair of coupled differential equations that describe the motion. Solve the equations of motion as discussed in Section 2.1.

7. Examine the motion of a pendulum with potential energy

$$V(\phi) = mgl \left(\frac{1}{2}\phi^2 - \frac{\phi^4}{24} \right).$$

   The potential energy is obtained by a Taylor expansion of the potential energy (2.15) around its minimum. The expansion for the potential serves as an approximation of the pendulum in cases where the displacements are no longer small.

8. Reconsider the approximaton made in the derivation of the period of the pendulum. Verify that the approximation used in equation (2.31) results in a further decrease of accuracy for larger angular displacements.

9. Examine the motion of the damped driven pendulum for different values of the external driving amplitude and for different damping constants. Study the changes in the Poincaré section for a certain set of initial conditions. Change the initial conditions and repeat the calculation

10. Simulate a vibrating string by using the procedures of Section 2.4.2.1. Examine the amplitudes and frequencies of the normal modes of vibration.

    Hint: Use a large number of springs.

11. Use the calculation presented in Section 2.5 to describe the motion of the moon around the earth. Do similar calculations for other pairs in the solar system.

12. Change the interaction potential of the central force in Section 2.5 to

$$V(r) = \frac{1}{|r_1 - r_2|^\alpha}.$$

    What type of motion results from this type of potential?

13. Show that the shortest distance between two points in (three-dimensional) space is a straight line.

14. Re–examine the Brachistochrone problem and show that the time required for a particle to move to the minimum point of the cycloid is $\pi/(a/g)^{1/2}$, independent of its starting point.

15. Derive the equations of motion for a rotating double pendulum.

16. Derive the equations of motion from the Lagrangian

$$L = \frac{i}{2}\hbar\rho_t - \frac{\hbar^2}{8m}\frac{\rho_x^2}{\rho} - \frac{1}{2m}\rho S_x^2 - \rho V(x) - \rho S_t \qquad (2.116)$$

    where the lower indices denote differentiations with respect to x and t. Compare the equations of motion with the Euler equations of hydrodynamics and the time-dependent Schrödinger equation of quantum mechanics. To represent the time-dependent Schrödinger equation, use the transformation $\psi = \rho^{1/2}\exp(iS/\hbar)$ where ρ and S are functions of x and t, respectively.

3

Electrodynamics

3.1 Potential and electric fields of discrete charge distributions

In electrostatic problems, we often need to determine the potential and the electric fields for a certain charge distribution. The basic equation of electrostatics is Gauss's law. From this fundamental relation connecting the charge density with the electric field, the potential of the field can be derived. We can state Gauss's law in differential form by

$$\operatorname{div} \boldsymbol{E} = 4\pi \varrho(\boldsymbol{r}). \tag{3.1}$$

If we introduce the potential Φ by $\boldsymbol{E} = -\operatorname{grad}\Phi$, we can rewrite (3.1) for a given charge distribution ϱ in the form of a Poisson equation

$$\Delta \Phi = -4\pi \varrho \tag{3.2}$$

where ϱ denotes the charge distribution. To obtain solutions of (3.2), we can use the Green's function formalism to derive a particular solution. The Green's function $G(\boldsymbol{r}, \boldsymbol{r}')$ itself has to satisfy a Poisson equation where the continuous charge density is replaced by Dirac's delta function $\Delta_r G(\boldsymbol{r}, \boldsymbol{r}') = -4\pi\delta(\boldsymbol{r} - \boldsymbol{r}')$. The potential Φ is then given by

$$\Phi(\boldsymbol{r}) = \int_V G(\boldsymbol{r}, \boldsymbol{r}')\varrho(\boldsymbol{r}')d^3r'. \tag{3.3}$$

In addition, we assume that the boundary condition $G|_{\partial V} = 0$ is satisfied on the surface of volume V. If the space in which our charges are located is infinitely extended, the Green's function is given in an analytic form by

$$G(\boldsymbol{r}, \boldsymbol{r}') = \frac{1}{|\boldsymbol{r} - \boldsymbol{r}'|}. \tag{3.4}$$

The solution of the Poisson equation (3.3) becomes

$$\Phi(\boldsymbol{r}) = \int \frac{\varrho(\boldsymbol{r}')}{|\boldsymbol{r} - \boldsymbol{r}'|} d^3 r'. \tag{3.5}$$

Our aim is to examine the potential and the electric fields of a discrete charge distribution. The charges are characterized by strength q_i and are located at certain positions $\boldsymbol{r}_i$. The charge density of such a distribution is given by

$$\varrho(\boldsymbol{r}) = \sum_{i=1}^{N} q_i \delta(\boldsymbol{r}_i). \tag{3.6}$$

The potential of such a discrete distribution of charges is in accordance with equation (3.5)

$$\Phi(\boldsymbol{r}) = \sum_{i=1}^{N} q_i \frac{1}{|\boldsymbol{r} - \boldsymbol{r}_i|}. \tag{3.7}$$

where r_i denotes the location of the point charge q_i. The corresponding electrical field is given by

$$\boldsymbol{E}(\boldsymbol{r}) = -\sum_{i=1}^{N} q_i \frac{\boldsymbol{r} - \boldsymbol{r}_i}{|\boldsymbol{r} - \boldsymbol{r}_i|^3} \tag{3.8}$$

and the energy density of the electric field of such a charge distribution is given by

$$w = \frac{1}{8\pi} |\boldsymbol{E}|^2. \tag{3.9}$$

Three fundamental properties of a discrete charge distribution are defined by equations (3.7), (3.8) and (3.9). In the following we write a *Mathematica* package which computes the potential, the electric field and the energy density for a given charge distribution. With this package, we are able to create pictures of the potential, the electric field and the energy density.

In order to design a graphical representation of the three quantities we need to create contour plots of a three-dimensional space. To simplify the handling of the functions, we enter the Cartesian coordinates of the locations and the strength of the charges as input variables in a list. Sublists of this list contain the information for specific charges. The structure of the input list is given by $\{\{x_1, y_1, z_1, \varrho_1\}, \{x_2, y_2, z_2, \varrho_2\}, \ldots\})$. To make things simple we choose the $y = 0$ section of the three-dimensional space. The package **PointCharge'** containing the equations discussed above generates contour plots of the potential, the electric field and the energy density. Following is a listing of the package.

```
                            ─────── File pointcha.m ───────
BeginPackage["PointCharge'"]

(* --- load additional standard packages --- *)

Needs["Graphics'PlotField'"]

Clear[Potential,Field,EnergyDensity,FieldPlot];

(* --- export functions --- *)

Potential::usage = "Potential[coordinates_List] creates the potential of
an assembly of point charges. The cartesian coordinates of the locations of
the charges are given in the form of {{x,y,z,charge},{x,y,z,charge},...}."

Field::usage = "Field[coordinates_List] calculates the electric field for
an ensemble of point charges. The cartesian coordinates are
lists in the form of {{x,y,z,charge},{...},...}."

EnergyDensity::usage = "EnergyDensity[coordinates_List] calculates the
density of the energy for an ensemble of point charges. The cartesian
coordinates are lists in the form of {{x,y,z,charge},{...},...}."

FieldPlot::usage = "FieldPlot[coordinates_List,typ_,options___] creates a
ContourPlot for an ensemble of point charges. The plot type (Potential,
Field, or Density) is specified as string in the second input variable. The
third argument allows a change of the Options of ContourPlot and
PlotGradientField."

(* --- define the global variables x,y,z  --- *)

x::usage
y::usage
z::usage

Begin["'Private'"]

(* --- determine the potential --- *)

Potential[coordinates_List]:=
      Block[{pot,x,y,z,ncharges},
      ncharges = Length[coordinates];
      pot = Sum[coordinates[[i,4]]/Sqrt[(x-coordinates[[i,1]])^2 +
                                        (y-coordinates[[i,2]])^2 +
                                        (z-coordinates[[i,3]])^2 ],
            {i,1,ncharges}]]

(* --- calculate the field ---*)

Field[coordinates_List]:=
      Block[{field,x,y,z,ncharges},
      ncharges = Length[coordinates];
      field = - Sum[coordinates[[i,4]]*({x,y,z}-{coordinates[[i,1]],
                                            coordinates[[i,2]],
                                            coordinates[[i,3]]})/
                          (Sqrt[(x-coordinates[[i,1]])^2 +
```

```
                                            (y-coordinates[[i,2]])^2 +
                                            (z-coordinates[[i,3]])^2 )^3,
              {i,1,ncharges}];
        Simplify[field]
        ]

(* --- calculate the energy --- *)

EnergyDensity[coordinates_List]:=
        Block[{density,x,y,z,field},
        field = Field[coordinates];
        density = field.field/(8*Pi)
        ]

(* --- create plots  --- *)

FieldPlot[coordinates_List,typ_,options___]:=
     Block[
      {pot, ncharges, xmin, xmax, zmin, zmax, xcoord = {}, zcoord = {},
      pl1, pl2},
       ncharges = Length[coordinates];
(* --- determine limits for the plot --- *)
         Do[
         AppendTo[xcoord,coordinates[[i,1]]];
         AppendTo[zcoord,coordinates[[i,3]]],
          {i,1,ncharges}];
       xmax = Max[xcoord]*1.5;
       zmax = Max[zcoord]*1.5;
       xmax = Max[{xmax,zmax}];
       zmax = xmax;
       xmin = -xmax;
       zmin = xmin;
       Clear[xcoord,zcoord];

(* --- fix the type of the plot ---*)
       If[typ == "Potential",pot = Potential[coordinates] /. y -> 0,
       If[typ == "Field",pot = -Potential[coordinates] /. y -> 0,
       If[typ == "EnergyDensity",pot = EnergyDensity[coordinates] /. y -> 0,
       Print[" "];
       Print["  wrong key word! Choose "];
       Print["  Potential, Field or EnergyDensity  "];
       Print["  to create a plot "];
       Return[]
       ]]];

(* --- plot the pictures --- *)
       If[typ == "Field",
         pl1 = PlotGradientField[pot,{x,xmin,xmax},{z,zmin,zmax},
             options,
             PlotPoints->20,
             ColorFunction->Hue
              ],
         pl1= ContourPlot[pot,{x,xmin,xmax},{z,zmin,zmax},
             options,
             PlotPoints->50,
             ColorFunction->Hue,
             Contours->15]
```

```
      ]
     ]
  End[]
  EndPackage[]
```

In order to test the functions of this package, we consider ensembles of charges frequently discussed in the literature. Our example is for two particles carrying the opposite charge. The input data list for the function **FieldPlot[ ]** is given by

```
  charges = {{1,0,0,1},{-1,0,0,-1}}
```

The charges are located in space at $x = 1, y = 0, z = 0$ and at $x = -1, y = 0, z = 0$. The fourth element in the sublists specifies the strength of the charges. The picture of the contour lines of the potential as given in figure 3.1 is created by calling

```
  FieldPlot[charges,"Potential"]
```

The second argument of **FieldPlot[ ]** is given as a string specifying the type of the contour plot. Possible values are *Potential*, *Field*, and *EnergyDensity*. A graphical representation of the energy density follows by

```
  FieldPlot[charges,"EnergyDensity"]
```

The result is given in figure 3.2.

3.2 Boundary problem of electrostatics

In the previous section, we discussed the arrangement of discrete charges. The problem was solved by means of the Poisson equation for the general case. We derived the solution for the potential using

$$\Delta\Phi = -4\pi\varrho. \tag{3.10}$$

Equation (3.10) is reduced to the Laplace equation if no charges are present in the space

$$\Delta\Phi = 0. \tag{3.11}$$

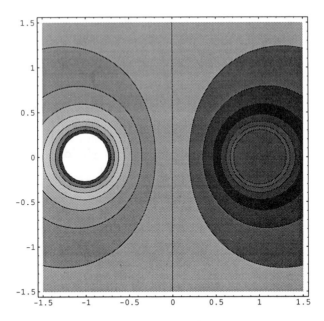

Figure 3.1.   Contour plot of the potential for two charges in the (x, z)–plane. The particles carry opposite charges.

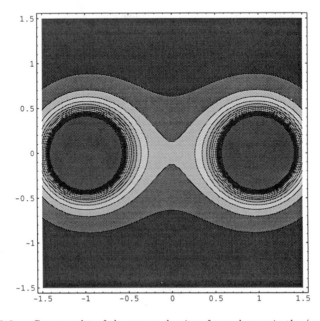

Figure 3.2.   Contour plot of the energy density of two charges in the (x, z)–plane.

The Laplace equation is a general type of equation applicable to many different theories in physics, such as continuum theory, gravitation, hydrodynamics, thermodynamics, and statistical physics. In this section we use both the Poisson and the Laplace equations (3.10) and (3.11) to describe electrostatic phenomena. We show that equations (3.10) and (3.11) are solvable by use of Green's function. If we know the Green's function of the equation, we are able to consider general boundary problems. A boundary problem is defined as follows: for a certain volume V, the surface of this volume, ∂V, possesses a specific electric potential. The question is to determine the electric potential inside the volume given the value on the surface. This type of electrostatic boundary problem is called a Dirichlet boundary value problem. According to equation (3.10) there are charges inside volume V. The distribution or density of these charges is denoted by $\varrho(\boldsymbol{x})$. The mathematical problem is to find solutions for equations (3.10) or (3.11) once we know the distribution of charges and the electric potential on the surface of the domain.

The Green's function allows us to simplify the solution of the problem. In our problem, we have to solve the Poisson equation (3.10) under certain restrictions. The Green's function related to the Poisson problem is defined by

$$\Delta G(\boldsymbol{x}, \boldsymbol{x}') = -4\pi\delta(\boldsymbol{x} - \boldsymbol{x}') \tag{3.12}$$

under the specific boundary condition

$$G(\boldsymbol{x}, \boldsymbol{x}')|_{\partial V} = 0 \qquad \text{with} \qquad \boldsymbol{x}' \in \partial V \tag{3.13}$$

on the surface ∂V of volume V.

In the previous section we discussed the Green's function for an infinitely extended space and found that the Green's function is represented by $G(\boldsymbol{r}, \boldsymbol{r}') = 1/|\boldsymbol{r} - \boldsymbol{r}'|$. The present problem is more complicated than the one previously discussed. We need to satisfy boundary conditions for a finite domain in space.

For our discussion we assume that the Green's function exists and that we can use it to solve the boundary problem. The proof of this assumption is given by [3.1]. The connection between the Green's function and the solution of the boundary problem is derived using Gauss's theorem. The first formula by Green

$$\int_V \operatorname{div} \boldsymbol{A} d^3x = \int_{\partial V} \boldsymbol{A} d^2\boldsymbol{f}, \tag{3.14}$$

along with an appropriate representation of the vector field $\boldsymbol{A} = \varPhi \cdot \nabla G - \nabla \varPhi \cdot G$ yields the second formula by Green

$$\operatorname{div} \boldsymbol{A} = \varPhi \Delta G - \Delta \varPhi G. \tag{3.15}$$

Using the integral theorem of Gauss in the form of equation (3.14) we find

$$\int_V (\varPhi \Delta G - \Delta \varPhi G) d^3x = \int_{\partial V} \left(\varPhi \frac{\partial G}{\partial n} - G \frac{\partial \varPhi}{\partial n} \right) d^2f, \tag{3.16}$$

where $\partial/\partial n = \boldsymbol{n} \cdot \nabla$ is the normal gradient. If we use relations (3.10), (3.12) and (3.13) in equation (3.16), we can derive the potential we are looking for

$$\begin{aligned} \Phi(\boldsymbol{x}) \quad &= \quad \int_V G(\boldsymbol{x}, \boldsymbol{x}')\varrho(\boldsymbol{x}')d^3x' \\ &\quad -\frac{1}{4\pi}\int_{\partial V} \Phi(\boldsymbol{x}')\frac{\partial G(\boldsymbol{x}, \boldsymbol{x}')}{\partial n'}d^2f'. \end{aligned} \qquad (3.17)$$

A comparison between equations (3.17) and (3.3) reveals that the total potential in the Dirichlet problem depends on a volume part (consistent with equation (3.3)) and on a surface part as well. The potential Φ at location $\boldsymbol{x}$ consists of a volume term containing the charges and of a surface term determined by the electric potential $\Phi(\boldsymbol{x}')$. The potential $\Phi(\boldsymbol{x}')$ used in the surface term is known as a boundary condition. If there are no charges in the present volume, the solution (3.17) reduces to

$$\Phi(\boldsymbol{x}) \quad = \quad -\frac{1}{4\pi}\int_{\partial V} \Phi(\boldsymbol{x}')\frac{\partial G(\boldsymbol{x}, \boldsymbol{x}')}{\partial n'}d^2f'. \qquad (3.18)$$

For the charge–free case, the electric potential at a location $\boldsymbol{x}$ inside the volume V is completely determined by the potential on the surface $\Phi(\boldsymbol{x}')$. We are able to derive equations (3.17) and (3.18) provided that the Green's function $G(\boldsymbol{x}, \boldsymbol{x}')$ vanishes on the surface of V. In other words, we assume the surface potential to be a boundary condition. This type of boundary condition is called a Dirichlet boundary condition. A second type is the so–called von Neumann boundary condition, which specifies the normal derivative of the electrostatic potential $\partial\Phi/\partial n$ on the surface. A third type used in potential theory is a mixture of Dirichlet and von Neumann boundary conditions. In the following we shall restrict ourselves to Dirichlet boundary conditions only.

If we take a closer look at solutions (3.17) and (3.18) of our boundary value problem, we observe that the Green's function as an unknown determines the solution of our problem. In other words, we solved the boundary problem in a form which contains an unknown function as defined by relation (3.12) and the boundary condition (3.13). The central problem is to find an explicit representation of the Green's function. One way to tackle this is by introducing an eigenfunction expansion [3.2]. This procedure always applies if the coordinates are separable. The eigenfunction expansion of the Green's function is based on the analogy between an eigenvalue problem and equations (3.10) and (3.11) for the potential.

The eigenvalue problem related to equation (3.10) is given by [3.2]

$$\Delta\Psi + (4\pi\varrho + \lambda)\Psi = 0. \qquad (3.19)$$

We assume that solutions Ψ of (3.19) satisfy the Dirichlet boundary conditions. In this case, the regular solutions of (3.19) only occur if parameter $\lambda = \lambda_n$ assumes certain discrete values. The λ_n's are the eigenvalues of equation (3.19). Their corresponding functions Ψ_n are eigenfunctions. The eigenfunctions Ψ_n are orthogonal and satisfy

$$\int_V \Psi_m^*(\boldsymbol{x})\Psi_n(\boldsymbol{x})d^3x = \delta_{mn}. \tag{3.20}$$

The eigenvalues of (3.19) can be discrete or continuous. In analogy to equation (3.12), the Green's function has to satisfy the equation

$$\Delta_x G(\boldsymbol{x}, \boldsymbol{x}') + \{4\pi\varrho + \lambda\}\, G(\boldsymbol{x}, \boldsymbol{x}') = -4\pi\delta(\boldsymbol{x} - \boldsymbol{x}'), \tag{3.21}$$

where λ is different to the eigenvalues λ_n. An expansion of the Green's function with respect to the eigenfunctions of the related eigenvalue problem is possible if the Green's function satisfies the same boundary conditions. Substituting an expansion of the Green's function

$$G(\boldsymbol{x}, \boldsymbol{x}') = \sum_n a_n(\boldsymbol{x}')\Psi_n(\boldsymbol{x}). \tag{3.22}$$

into equation (3.21), we get

$$\sum_m a_m(\boldsymbol{x}')(\lambda - \lambda_m)\Psi_m(\boldsymbol{x}) = -4\pi\delta(\boldsymbol{x} - \boldsymbol{x}'). \tag{3.23}$$

Multiplying both sides of equation (3.23) by $\Psi_n^*(\boldsymbol{x})$ and integrating the result over the entire volume, we obtain the expansion coefficients $a_m(\boldsymbol{x}')$. Using the orthogonal relation (3.20) simplifies the sum. The expansion coefficients are defined by

$$a_n(\boldsymbol{x}') = 4\pi\frac{\Psi_n^*(\boldsymbol{x}')}{\lambda_n - \lambda}. \tag{3.24}$$

With relation (3.24) we get the representation of the Green's function

$$G(\boldsymbol{x}, \boldsymbol{x}') = 4\pi \sum_n \frac{\Psi_n^*(\boldsymbol{x}')\Psi_n(\boldsymbol{x})}{\lambda_n - \lambda}. \tag{3.25}$$

So far our considerations have assumed a discrete spectrum of eigenvalues. For a continuous distribution of eigenvalues λ_n, we need to replace the sum in equation (3.25) with an integral over the eigenvalues.

By using the representation of the Green's function (3.25), we can rewrite the solution of the potential (3.17) and (3.18) in the form

$$
\begin{aligned}
\Phi(\boldsymbol{x}) &= \int_V 4\pi \sum_n \frac{\Psi_n^*(\boldsymbol{x}')\Psi_n(\boldsymbol{x})}{\lambda_n - \lambda}\varrho(\boldsymbol{x}')d^3x' \\
&\quad - \int_{\partial V} \Phi(\boldsymbol{x}') \sum_n \frac{\Psi_n(\boldsymbol{x})}{\lambda_n - \lambda}\frac{\partial \Psi_n^*(\boldsymbol{x}')}{\partial n'}d^2f' \\
&= 4\pi \sum_n \frac{\Psi_n(\boldsymbol{x})}{\lambda_n - \lambda}\int_V \Psi_n(\boldsymbol{x}')\varrho(\boldsymbol{x}')d^3x' \\
&\quad - \sum_n \frac{\Psi_n(\boldsymbol{x})}{\lambda_n - \lambda}\int_{\partial V} \Phi(\boldsymbol{x}')\frac{\partial \Psi_n^*(\boldsymbol{x}')}{\partial n'}d^2f'.
\end{aligned} \tag{3.26}
$$

If we know the eigenfunctions and eigenvalues of the problem, we can represent the potential by

$$\Phi(\boldsymbol{x}) = \sum_n (c_n - d_n)\Psi_n(\boldsymbol{x}) \tag{3.27}$$

where the c_n's and the d_n's are expansion coefficients defined by

$$c_n = \frac{4\pi}{\lambda_n - \lambda} \int_V \Psi_n^*(\boldsymbol{x}')\varrho(\boldsymbol{x}')d^3x' \tag{3.28}$$

and

$$d_n = \frac{1}{\lambda_n - \lambda} \int_{\partial V} \Phi(\boldsymbol{x}') \frac{\partial \Psi_n^*(\boldsymbol{x}')}{\partial n'} d^2 f'. \tag{3.29}$$

For the charge–free case $\varrho = 0$, we find

$$\Phi(\boldsymbol{x}) = -\sum_n \frac{\Psi_n(\boldsymbol{x})}{\lambda_n - \lambda} \int_{\partial V} \Phi(\boldsymbol{x}') \frac{\partial \Psi_n^*(\boldsymbol{x}')}{\partial n'} d^2 f'. \tag{3.30}$$

which reduces to

$$\Phi(\boldsymbol{x}) = -\sum_n d_n \Psi_n(\boldsymbol{x}). \tag{3.31}$$

The unknown quantities of this representation are the eigenfunctions Ψ_n and the expansion coefficients c_n and d_n. By examining a specific planar problem we show how these unknowns are calculated. To make things simple, we assume that no charges are distributed on the plane.

The problem under consideration examines in a section of a disk in which boundaries have fixed potential values $\Phi(r, \phi = 0) = 0$, $\Phi(r, \phi = \alpha) = 0$, and $\Phi(r = R, \phi) = \Phi_0(\phi)$. The specific form of the domain and the boundary values are given in figure 3.3.

The domain G is free of any charges and the potential $\Phi(r, \phi)$ is regular and finite for $r \to 0$. To solve the problem efficiently, we choose coordinates which reflect the geometry of our problem. In this case they are plane cylindrical coordinates. Since G is free of any charges, Laplace's equation in plane cylindrical coordinates takes the form

$$\frac{1}{r}\frac{\partial}{\partial r}\left(r\frac{\partial \Phi}{\partial r}\right) + \frac{1}{r^2}\frac{\partial^2 \Phi}{\partial \phi^2} = 0. \tag{3.32}$$

When deriving the solution, we assume that the coordinates are separated. If we use the assumption of separating the coordinates as we have done in our example, we are able to express the electric potential as $\Phi(r, \phi) = g(r)h(\phi)$. Substituting this expression into equation (3.32), we get

$$\frac{r}{g(r)}\frac{d}{dr}\left(r\frac{dg}{dr}\right) = -\frac{1}{h(\phi)}\frac{d^2 h(\phi)}{d\phi^2} = \nu^2 \tag{3.33}$$

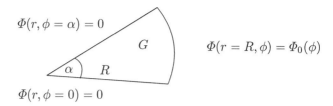

Figure 3.3. Boundary conditions on a disk segment. The domain G is free of charges.

where ν is a constant. Separating both equations, we get two ordinary differential equations determining g and h. g and h represent the eigenfunctions of the Green's function

$$\frac{r}{g(r)}\frac{d}{dr}\left(r\frac{dg}{dr}\right) = \nu^2 \tag{3.34}$$

$$\frac{1}{h(\phi)}\frac{d^2h}{d\phi^2} = -\nu^2. \tag{3.35}$$

The eigenfunctions of the radial part of the potential are

$$g_\nu(r) = a_\nu r^\nu + b_\nu r^{-\nu}. \tag{3.36}$$

The angular part of the eigenfunctions defined in equation (3.34) is given by

$$h_\nu(\phi) = A_\nu \sin(\nu\phi) + B_\nu \cos(\nu\phi). \tag{3.37}$$

The solutions (3.36) and (3.37) contain four constants a_ν, b_ν, A_ν, and B_ν for each eigenvalue ν. These constants have to satisfy the boundary conditions and the condition of regularity at $r = 0$.

Let us first examine the radial part of the solution in the domain G. We find that for $\phi = 0$ the relation

$$\Phi(r, \phi = 0) = g(r)h(\phi = 0) = 0 \tag{3.38}$$

needs to be satisfied. From condition (3.38) it follows that $h(\phi = 0) = B_\nu = 0$. From the boundary condition at $\phi = \alpha$ we get the condition

$$\Phi(r, \phi = \alpha) = g(r)h(\phi = \alpha) = 0 \tag{3.39}$$

which results in $h(\alpha) = A_\nu \sin(\nu\alpha) = 0$. As a consequence we get $\nu = n\pi/\alpha$ with $n = 0, 1, 2, 3, \ldots$. The angular part of the solution thus reduces to

$$h_n(\phi) = A_n \sin\left(\frac{n\pi}{\alpha}\phi\right). \tag{3.40}$$

From the condition of regularity $\Phi(r \to 0, \phi) < \infty$ it follows from

$$\Phi(r, \phi) = h_\nu(\phi)(a_\nu r^\nu + b_\nu r^{-\nu}) \tag{3.41}$$

that $b_\nu = 0$. The solution of the potential is thus represented by

$$\Phi(r, \phi) = \sum_{n=0}^{\infty} d_n r^{n\pi/\alpha} \sin\left(\frac{n\pi}{\alpha}\phi\right) \tag{3.42}$$

where $d_n = a_n A_n$. Expression (3.42) contains the unknown coefficients d_n which we need to determine in order to find their explicit representations. Values for d_n are determined by applying the boundary condition on the circle $\Phi(r = R, \phi) = \Phi_0(\phi)$. If we take into account the orthogonality relation for the trigonometric functions

$$\frac{2}{\alpha} \int_0^\alpha d\phi \sin\left(\frac{n\pi}{\alpha}\phi\right) \sin\left(\frac{m\pi}{\alpha}\phi\right) = \delta_{mn}, \tag{3.43}$$

we are able to derive from the boundary condition of the circle a representation of d_n by

$$\int_0^\alpha d\phi\Phi_0(\phi) \sin\left(\frac{m\pi}{\alpha}\phi\right) = \sum_{m=0}^{\infty} d_m R^{m\pi/\alpha} \underbrace{\int_0^\alpha d\phi \sin\left(\frac{n\pi}{\alpha}\phi\right) \sin\left(\frac{m\pi}{\alpha}\phi\right)}_{\frac{\alpha}{2}\delta_{nm}}$$

$$= \frac{\alpha}{2} R^{n\pi/\alpha} d_n \tag{3.44}$$

or in explicit form

$$d_n = R^{-n\pi/\alpha} \frac{2}{\alpha} \int_0^\alpha d\phi\Phi_0(\phi) \sin\left(\frac{n\pi}{\alpha}\phi\right). \tag{3.45}$$

The representation of d_n by the integral (3.45) includes the boundary condition and only contains known parameters. Thus we can determine d_n's numerical value if we know the boundary condition and if we specify the index m of the expansion in (3.42). The values of d_n are, however, only defined if the integral in (3.45) converges. The specific form of the Green's function is derivable if we compare the representation of the solution (3.42) with the definition of the Green's function.

With the above theoretical considerations, an explicit representation of the solution is now necessary. By specifying the geometrical parameters of the problem, the radius R of the segment and the angle α, the potential value along the rim of the disk and equation (3.42), we can calculate the potential in the domain G. The central quantities of the expansion (3.42) are the coefficients d_n. In order to make these factors available in *Mathematica*, we define the sum (3.42) and the integral (3.45) in the **Potential[]** function of the package **BoundaryProblem'**. We define relations (3.42) and (3.45) to control the accuracy of the calculation using an upper summation index n (see also the definition of the *Mathematica* function **Potential[]** in the file **boundary.m**). A listing of the package is given below. An example of the potential for the parameters $R = 1$, $\alpha = \pi/4$, and $\Phi_0(\phi) = 1$ is given in figure 3.4. The calling sequence of **Potential[]** takes the form **Potential[f[x],R,alpha,n]**.

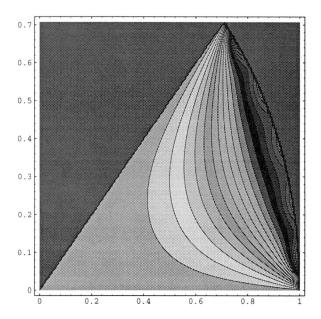

Figure 3.4.   Contour plot of the potential in the domain G. Boundary conditions and geometric parameters are $\Phi_0(\phi) = 1$, $R = 1$, $\alpha = \pi/4$ and $n = 10$. The calling sequence to create this picture is **Potential[1,1,Pi/4,10]** .

```
——————————————————— File boundary.m ———————————————————

BeginPackage["BoundaryProblem'"]

Clear[Potential,pot1]

Potential::usage = "Potential[boundary_,R_,alpha_,n_] calculates the
potential in a circular segment. Input parameters are the potential on the
circle, the radius R of the circle and the angle of the segment of the circle.
The last argument n determines the number of expansion terms used to
represent the solution."

Begin["'Private'"]

Potential[boundary_,R_,alpha_,n_] := Block[{listed = {}},
(* --- replace the independent variable in the input by Phi --- *)
      boundaryh = boundary /. f_[x2_.*x1_] -> f[x2*phi];
(* --- calculate the coefficients of the expansion d_n --- *)
      Do[
         AppendTo[listed,
               Integrate[boundaryh*Sin[m*Pi*phi/alpha], {phi,0,alpha}]*
               R^(m*Pi/alpha)*2/alpha],
          {m,0,n}];
(* --- calculate the potential with the sum --- *)
      pot = Sum[listed[[n1+1]]*r^(n1*Pi/alpha)*Sin[n1*Pi*phi/alpha],
          {n1,0,n}];
(* --- transform the potential to cartesian coordinates --- *)
```

```
(* --- Which[] allows the definition of alternative values --- *)
    pot1[x_,y_] := N[Which[x != 0 && Sqrt[x^2+y^2] <= R &&
                         ArcTan[y/x] <= alpha,
              poth = pot /. {r->Sqrt[x^2+y^2],phi->ArcTan[y/x]},
                         x == 0 && y == 0,
              poth = pot /. {r->0},
                         x == 0 && y > 0,
              poth = -0.1,
                         Sqrt[x^2+y^2] > R || ArcTan[y/x] > alpha,
              poth = -0.1
                     ]];
(* --- graphical representation of the potential by ContourPlot --- *)
    ContourPlot[pot1[x,y],{x,0,R},
                         {y,0,R*Sin[alpha]},
              ColorFunction->Hue,
              Contours->20,
              PlotPoints->200]
     ]
End[]
EndPackage[]
```

3.3   Two ions in the Penning trap

The study of spectroscopic properties of single ions requires that one or two ions are available. Nowadays, ions can be successfully separated and stored by means of ion traps. Two techniques are used for trapping ions. The first method uses a dynamic electric field, while the second method uses static electric and magnetic fields. The dynamic trap was originally invented by Paul [3.3]. The static trap is based on the work of Penning [3.4]. Both traps use a combination of electric and magnetic fields to confine ions in a certain volume of space. Two paraboloids connected to a *dc* source determine the kind of electric field in which the ions are trapped. The form of the paraboloids in turn determines the field of the trap's interior. Since the motion of the ions in Paul's trap is very complicated, we restrict our study to the Penning trap.

In our discussion of the Penning trap, the form of the quadrupole fields determined by the shapes of the paraboloids is assumed to be

$$\Phi = \frac{U_0}{r_0^2 + 2z_0^2}(x^2 + y^2 - 2z^2) \tag{3.46}$$

where U_0 is the strength of the source and r_0 and z_0 are the radial and axial extensions of the trap (see figure 3.5). The shape of the potential is a consequence of the Laplace equation $\Delta\Phi = 0$. The given functional shape of the potential is experimentally created by conducting walls which are connected to a dc battery. The force acting on an ion carrying charge q in the trap is given by

$$\boldsymbol{F} = q\boldsymbol{E} = -q\nabla\Phi. \tag{3.47}$$

145

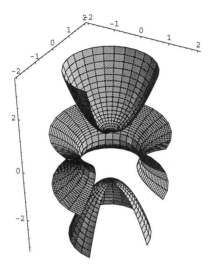

Figure 3.5. Cross-section of the Penning trap. The paraboloids are positioned on dc potentials. A constant magnetic field is superimposed in the vertical direction (not shown). The ions move in the center of the trap.

From the functional form of the electric field $\boldsymbol{E}$ of the trap

$$\boldsymbol{E} = -\nabla\Phi = -\frac{2U_0}{r_0^2 + 2z_0^2}\begin{pmatrix} x \\ y \\ -2z \end{pmatrix} = -\frac{2U_0}{r_0^2 + 2z_0^2}(\boldsymbol{x} - 3z\boldsymbol{e}_z), \qquad (3.48)$$

we detect a change of sign in the coordinates. The different signs in x, y, and z direction causes an instability along the z–axis. This instability allows the ions to escape the trap. To prevent escape from the trap in the z–direction, Paul and co-workers used a high frequency ac field while Penning and co-workers used a permanent magnetic field $\boldsymbol{B} = B_0\boldsymbol{e}_z$.

In a static trap the forces acting on each of the two ions are determined by the electromagnetic force of the external fields and the repulsive force of the Coulomb interaction of the charges. The external fields consist of the static magnetic field along the z–axis and the electric quadrupole field of the trap. The Coulomb interaction of the two particles is mainly governed by the charges which are carried by the particles. The total force on each particle is a combination of trap and Coulomb forces. Since we have a system containing only a few particles, we can use Newton's theory to write down the equations of motion in the form

$$m\ddot{\boldsymbol{x}}_i = \boldsymbol{F}_i^{(T)} + \boldsymbol{F}_i^{(Coul)} \qquad i = 1, 2. \qquad (3.49)$$

In equation (3.49) $F_i^{(T)}$ the trap force denotes the Lorentz force of a particle in the electromagnetic field given by

$$F_i^{(T)} = qE_i + q(v_i \times B). \tag{3.50}$$

Since the magnetic field B is a constant field along the z–direction

$$B = B_0 e_z, \tag{3.51}$$

the total trap force on the i-th ion is given by

$$F_i^{(T)} = -\frac{2U_0 q}{r_0^2 + 2z_0^2}(x_i - 3z_i e_z) + q(\dot{x}_i \times B). \tag{3.52}$$

The Coulomb forces between the first and the second ion are

$$F_{12}^{(Coul)} = \frac{q^2}{4\pi\epsilon_0}\frac{x_1 - x_2}{|x_1 - x_2|^3} \tag{3.53}$$

$$F_{21}^{(Coul)} = \frac{q^2}{4\pi\epsilon_0}\frac{x_2 - x_1}{|x_1 - x_2|^3}. \tag{3.54}$$

The explicit form of the equations of motion are thus

$$m\ddot{x}_1 = -\frac{2U_0 q}{r_0^2 + 2z_0^2}(x_1 - 3z_1 e_z) + q(\dot{x}_1 \times B) + \frac{q^2}{4\pi\epsilon_0}\frac{x_1 - x_2}{|x_1 - x_2|^3} \tag{3.55}$$

$$m\ddot{x}_2 = -\frac{2U_0 q}{r_0^2 + 2z_0^2}(x_2 - 3z_2 e_z) + q(\dot{x}_2 \times B) + \frac{q^2}{4\pi\epsilon_0}\frac{x_2 - x_1}{|x_1 - x_2|^3}. \tag{3.56}$$

The two equations of motion (3.55) and (3.56) are coupled ordinary differential equations of the second order. They can be decoupled by introducing relative and center of mass coordinates

$$r = x_1 - x_2 \tag{3.57}$$

$$R = \frac{1}{2}(x_1 + x_2).$$

Using equations (3.57) in (3.55) and (3.56), we can describe the motion of the two ions in center of mass and in relative coordinates. The two decoupled motions describe a quasi relative particle and the center of mass. The two transformed equations read

$$\ddot{R} = -\frac{2U_0 q}{m(r_0^2 + 2z_0^2)}(R - 3Z e_z) + \frac{B_0 q}{m}(\dot{R} \times e_z) \tag{3.58}$$

$$\ddot{r} = -\frac{2U_0 q}{m(r_0^2 + 2z_0^2)}(r - 3z e_z) + \frac{B_0 q}{m}(\dot{r} \times e_z) + \frac{2q^2}{4\pi m\epsilon_0}\frac{r}{|r|^3}. \tag{3.59}$$

If we assume that the two ions carry a negative charge $q < 0$ and that the dc potential U_0 on the paraboloids is positive ($U_0 > 0$), then we can introduce two characteristic frequencies and a scaled charge by

$$w_0^2 = \frac{2U_0|q|}{m(r_0^2 + 2z_0^2)} \tag{3.60}$$

$$\omega_c = \frac{|q|B_0}{m} \tag{3.61}$$

$$Q^2 = \frac{2q^2}{4\pi\epsilon_0 m}. \tag{3.62}$$

Constant w_0 denotes the frequency of the oscillations along the z–direction. ω_c is the cyclotron frequency (i.e., the frequency with which the ions spin around the magnetic field). Q represents the scaled charge. Using these constants in the equations of motion (3.58) and (3.59) we get a simplified system of equations containing only three constants

$$\ddot{\boldsymbol{R}} = w_0^2(\boldsymbol{R} - 3Z\boldsymbol{e}_Z) - \omega_c(\dot{\boldsymbol{R}} \times \boldsymbol{e}_Z) \tag{3.63}$$

$$\ddot{\boldsymbol{r}} = w_0^2(\boldsymbol{r} - 3z\boldsymbol{e}_z) - \omega_c(\dot{\boldsymbol{r}} \times \boldsymbol{e}_z) + \frac{Q^2\boldsymbol{r}}{|\boldsymbol{r}|^3}. \tag{3.64}$$

In the following subsections we discuss the two different types of motion resulting from these equations.

3.3.1 The center of mass motion

The center of mass motion is determined by equation (3.63). Writing down the equations of motion in Cartesian coordinates X, Y, and Z, we get a coupled system of equations

$$\ddot{X} - w_0^2 X + \omega_c \dot{Y} = 0 \tag{3.65}$$
$$\ddot{Y} - w_0^2 Y - \omega_c \dot{X} = 0 \tag{3.66}$$
$$\ddot{Z} + 2w_0^2 Z = 0. \tag{3.67}$$

The equations of motion for the X and Y components are coupled through the cross product. The Z component of the motion is completely decoupled from the X and Y coordinates. The last of these three equations is equivalent to a harmonic oscillator with frequency $\sqrt{2}w_0$. Thus we immediately know the solution of the Z coordinate given by

$$Z(t) = A\cos(\sqrt{2}w_0 t + B). \tag{3.68}$$

The arbitrary constants A and B are related to the initial conditions of the motion by $Z(t = 0) = Z_0$ and $\dot{Z}(t = 0) = \dot{Z}_0$. Therefore $A = Z_0^2 + \dot{Z}_0^2/(2w_0^2)$ and $\tan B = \dot{Z}_0/(\sqrt{2}w_0 Z_0)$.

A representation of the solution of the remaining two equations (3.65) and (3.66) follows if we combine the two coordinates X and Y by a complex transformation of the form $\Upsilon = X + iY$. Applying this transformation to the two equations gives us the simple representation

$$\ddot{\Upsilon} - w_0^2 \Upsilon - i\omega_c \dot{\Upsilon} = 0. \tag{3.69}$$

If we assume that the solutions of equation (3.69) are harmonic functions of the type $\Upsilon = e^{i\omega t}$, we get the corresponding characteristic polynomial

$$\omega(\omega_c - \omega) - \omega_0^2 = 0. \tag{3.70}$$

The two solutions of this quadratic equation are given by the frequencies ω_1 and ω_2

$$\omega_1 = \frac{\omega_c}{2} + \sqrt{\left(\frac{\omega_c}{2}\right)^2 - \omega_0^2} \tag{3.71}$$

$$\omega_2 = \frac{\omega_c}{2} - \sqrt{\left(\frac{\omega_c}{2}\right)^2 - \omega_0^2}. \tag{3.72}$$

The two frequencies are combinations of the cyclotron frequency ω_c and the axial frequency ω_0. The general solution of equations (3.65) and (3.66) is thus given by

$$\begin{aligned} X(t) &= B_r \cos(\omega_1 t) + B_i \sin(\omega_1 t) \\ &\quad + A_r \cos(\omega_2 t) + A_i \sin(\omega_2 t) \end{aligned} \tag{3.73}$$

$$\begin{aligned} Y(t) &= A_r \sin(\omega_2 t) - A_i \cos(\omega_2 t) \\ &\quad + B_r \sin(\omega_1 t) - B_i \cos(\omega_1 t). \end{aligned} \tag{3.74}$$

The constants of integration A_r, A_i, B_r and B_i are related to the initial conditions $X_0, Y_0, \dot{X}_0$ and $\dot{Y}_0$ by the relations

$$A_r = \frac{\dot{Y}_0 - \omega_1 X_0}{\omega_2 - \omega_1} \tag{3.75}$$

$$A_i = \frac{\dot{X}_0 + \omega_1 Y_0}{\omega_2 - \omega_1} \tag{3.76}$$

$$B_r = \frac{\dot{Y}_0 - \omega_2 X_0}{\omega_1 - \omega_2} \tag{3.77}$$

$$B_i = \frac{\dot{X}_0 + \omega_2 Y_0}{\omega_1 - \omega_2}. \tag{3.78}$$

A special case of solutions (3.73) and (3.74) is obtained if we assume that the center of mass is initially located in the origin of the coordinate system $X_0 = Y_0 = 0$. We get from (3.75) $A_r = -B_r$, and $A_i = -B_i$. The solution then takes the form

$$\begin{aligned} X(t) &= A_r \sin\left(\frac{\omega_c}{2}t\right) \sin\left(\sqrt{\left(\frac{\omega_c}{2}\right)^2 - \omega_0^2}\, t\right) \\ &\quad - A_i \cos\left(\frac{\omega_c}{2}t\right) \sin\left(\sqrt{\left(\frac{\omega_c}{2}\right)^2 - \omega_0^2}\, t\right) \end{aligned} \tag{3.79}$$

$$\begin{aligned} Y(t) &= A_i \sin\left(\frac{\omega_c}{2}t\right) \sin\left(\sqrt{\left(\frac{\omega_c}{2}\right)^2 - \omega_0^2}\, t\right) \\ &\quad - A_r \cos\left(\frac{\omega_c}{2}t\right) \sin\left(\sqrt{\left(\frac{\omega_c}{2}\right)^2 - \omega_0^2}\, t\right). \end{aligned} \tag{3.80}$$

The above solutions show that the motion of the center of mass in the X, Y plane is governed by two frequencies. The first frequency is one half of the cyclotron frequency ω_c while the second frequency is a combination of the axial frequency and the cyclotron frequency given by $\sqrt{(\omega_c/2)^2 - \omega_0^2}$. A plot of the motion in center of mass coordinates is given in figure 3.6. The three-dimensional motion of the center of mass is governed by three frequencies. The axial frequency $\sqrt{2}\omega_0$ determines the oscillation rate of the center of mass along the z–axis. The halved cyclotron frequency $\omega_c/2$ governs the spinning of the particles around the magnetic lines.

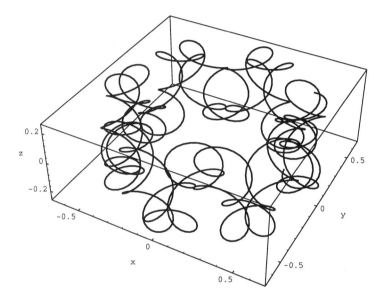

Figure 3.6.   Motion of the center of mass in space for $t \in [0, 100]$. The initial conditions are $X_0 = 0.5 = Y_0$, $\dot{X}_0 = 0.1 = \dot{Y}_0$. The cyclotron frequency is fixed at $\omega_c = 5$.

3.3.2 Relative motion of the ions

The relative motion of the two ions is governed by equation (3.64)

$$\ddot{\boldsymbol{r}} - \omega_0^2(\boldsymbol{r} - 3z\boldsymbol{e}_z) + \omega_c(\dot{\boldsymbol{r}} \times \boldsymbol{e}_z) = \frac{Q^2\boldsymbol{r}}{|\boldsymbol{r}|^3}. \tag{3.81}$$

Cylindrical coordinates are the appropriate coordinate system giving an efficient description of the relative motion of the particles. Location $\boldsymbol{r}$ of the relative particle is given in cylindrical coordinates by the representation

$$\boldsymbol{r} = \varrho\boldsymbol{e}_\varrho + \zeta\boldsymbol{e}_z \tag{3.82}$$

where e_ϱ and e_z represent the unit vectors in the radial and axial directions. Using these coordinates in the equation of motion (3.81) gives the following representation

$$(\ddot{\varrho} - \varrho\dot{\phi}^2)e_\varrho + (2\dot{\varrho}\dot{\phi} + \varrho\ddot{\phi})e_\phi + \ddot{\zeta}e_z \tag{3.83}$$

$$-\omega_0^2(\varrho e_\varrho - 2\zeta e_z) + \omega_c(-\dot{\varrho}e_\phi + \varrho\dot{\phi}e_\varrho) = \frac{Q^2\varrho e_\varrho + \zeta e_z}{(\varrho^2 + \zeta^2)^{3/2}}.$$

Separating each coordinate direction, we can split equation (3.83) into a system of equations for the coordinates ϱ, ϕ, and ζ

$$\ddot{\varrho} - \varrho\dot{\phi}^2 - \omega_0^2\varrho + \omega_c\varrho\dot{\phi} = \frac{Q^2\varrho}{(\varrho^2 + \zeta^2)^{3/2}} \tag{3.84}$$

$$2\dot{\varrho}\dot{\phi} + \varrho\ddot{\phi} - \omega_c\dot{\varrho} = 0 \tag{3.85}$$

$$\ddot{\zeta} + 2\omega_0^2\zeta = \frac{Q^2\zeta}{(\varrho^2 + \zeta^2)^{3/2}}. \tag{3.86}$$

By multiplying equation (3.85) by the radial coordinate ϱ and integrating the result, we are able to derive an integral of motion. This integral of motion is given by an extended angular momentum containing the cyclotron frequency and is thus connected with the magnetic field. The conserved quantity is given by

$$\ell_B = \varrho^2\dot{\phi} - \frac{\omega_c}{2}\varrho^2. \tag{3.87}$$

The integral of motion (3.87) eliminates the ϕ dependence in equation (3.84). The elimination of ϕ reduces the system of equations (3.84) and (3.86) to

$$\ddot{\varrho} + \left(\left(\frac{\omega_c}{2}\right)^2 - \omega_0^2\right)\varrho - \frac{\ell_B^2}{\varrho^3} = \frac{Q^2\varrho}{(\varrho^2 + \zeta^2)^{3/2}} \tag{3.88}$$

$$\ddot{\zeta} + 2\omega_0^2\zeta = \frac{Q^2\zeta}{(\varrho^2 + \zeta^2)^{3/2}}. \tag{3.89}$$

This system of equations contains a multitude of parameters. Our aim is to reduce these parameters by appropriately scaling the temporal and spatial coordinates. If we consider the expression $\beta = (\omega_c/2)^2 - \omega_0^2 > 0$ to be positive, time is scaled by $\tau = \beta t$. The radial and axial coordinates ϱ and ζ are scaled by the factor $d = (Q/\beta)^{2/3}$. Introducing the abbreviations $\nu^2 = (\ell_B/\beta)^2$ and $\lambda^2 = (\sqrt{2}\omega_0^2/\beta)^2$ simplifies the system of equations (3.88) and (3.89) to

$$\ddot{\varrho} + \varrho - \frac{\nu^2}{\varrho^3} = \frac{\varrho}{(\varrho^2 + \zeta^2)^{3/2}} \tag{3.90}$$

$$\ddot{\zeta} + \lambda^2\zeta = \frac{\zeta}{(\varrho^2 + \zeta^2)^{3/2}}, \tag{3.91}$$

containing only two parameters ν and λ. The handling of equations (3.90) and (3.91) is easier than the four parameter representation in equations (3.88) and (3.89). Note that equations (3.90) and (3.91) are equivalent to the secular equations of the Paul trap. Both systems of equations are derived from a Lagrangian given by

$$\mathcal{L} = \frac{1}{2}(\dot{\varrho}^2 + \dot{\zeta}^2) - \left\{ \frac{1}{2}(\varrho^2 + 2\lambda^2\zeta^2) + \frac{1}{(\varrho^2 + \zeta^2)^{1/2}} + \frac{\nu^2}{2\varrho^2} \right\}. \qquad (3.92)$$

Equations (3.90) and (3.91) form a highly nonlinear coupled system of equations which can only be solved analytically given a special choice of parameters λ and ν [3.5]. If we wish to choose parameters, we need to integrate the equations numerically. *Mathematica* supports numerical integrations and we use this property to find numerical solutions for equations (3.90) and (3.91). The package **Penning'** contains the necessary function **PenningI[]** to integrate equations (3.90) and (3.91). Function **PenningI[]** also provides several graphical representations of the potential, the path of the relative particle, and a superposition of the two graphics. Since the equations of motion are equivalent to a Hamiltonian system, **PenningI[]** first plots the effective potential in which the motion takes place. In a second step, **PenningI[]** creates a graphical representation of the path in the variables ϱ and ζ. The last output of **PenningI[]** is a graphic in which the two previously produced pictures are superimposed. An example of a typical path in the potential is given in figure 3.7. Parameters λ and ν of this figure have been chosen so that the motion of the relative particle is regular. Figure 3.8 shows a path for parameters λ and ν where chaotic motion is present.

File penning.m

```
BeginPackage["Penning'"]

Clear[V,PenningI,PenningCMPlot]

PenningI::usage = "PenningI[r0_,z0_,e0_,n_,l_,te_] determines the numerical
solution of the equation of motion for the relative components. To integrate
the equations of motion, the initial conditions r0 = r(t=0), z0 = z(t=0) and
the total energy e0 are needed as input parameters. The momentum with respect
to the r direction is set to pr0=0. Parameters l and n determine the
shape of the potential. The last argument te specifies the end point of
the integration."

PenningCMPlot::usage = "PenningCMPlot[x0_,y0_,x0d_,y0d_,w_] gives a graphical
representation of the center of mass motion for two ions in the Penning trap.
The plot is created for a fixed cyclotron frequency w in cartesian
coordinates (x,y,z). x0, y0, x0d, and y0d are the initial conditions for
integration."

Begin["'Private'"]

(* --- potential --- *)
V[x_, y_, l_, n_] := (x^2 + l^2*y^2)/2 + n^2/(2*x^2) +
                     1/(x^2 + y^2)^(1/2)

(*--- numerical integration of the relative motion ---*)

PenningI[r0_,z0_,e0_,n_,l_,te_]:=Module[{intk,pz0},
```

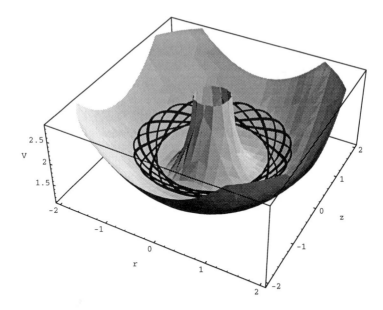

Figure 3.7.   Relative motion in a Penning trap for $\lambda = 1$ and $\nu = 0$. The plot of the particle is superimposed on the effective potential. The numerical integration extends over $t \in [0, 100]$. The initial conditions are $\varrho_0 = 1.1$, $\zeta_0 = 0.5$, $\dot{\varrho}_0 = 0.0$, and $E = 2.0$.

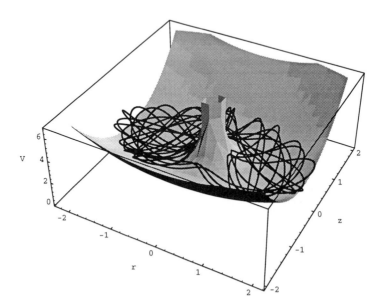

Figure 3.8.   Relative motion in a Penning trap for $\lambda = 1.75$ and $\nu = 0$. The plot of the particle is superimposed on the effective potential. The numerical integration extends over $t \in [0, 100]$. Initial conditions are $\varrho_0 = 1.0$, $\zeta_0 = 0.0$, $\dot{\varrho}_0 = 0.0$, and $E = 3.0$.

```
(* --- initial value of the momentum in z direction --- *)
        pz0 = Sqrt[2*(e0-V[r0,z0,1,n])];
(* --- numerical solution of the initial value problem --- *)
intk = NDSolve[{pr'[t] == n^2/r[t]^3 - r[t] +
                        r[t]/(r[t]^2+z[t]^2)^(3/2),
               pz'[t] == -1^2*z[t] + z[t]/(r[t]^2+z[t]^2)^(3/2),
               r'[t]  == pr[t],
               z'[t]  == pz[t],
(* --- initial values --- *)
        r[0] == r0, z[0] == z0, pr[0] == 0, pz[0] == pz0},
        {r,z,pr,pz},{t,0,te}, MaxSteps->3000];
(* --- graphical representation --- *)
(* --- save the plots in the file penning.ps --- *)
Display["penning.ps",
(* --- plot the potential --- *)
Show[{Plot3D[V[x,y,1,n]-0.4,{x,-2,2},{y,-2,2},Mesh->False,
            PlotPoints->25],
(* --- plot the tracks by ParametricPlot3D --- *)
     ParametricPlot3D[Evaluate[{r[t],z[t],V[r[t],z[t],1,n]} /. intk],
                     {t,0,te},PlotPoints->1000,
                                    AxesLabel->{r,z,V}]},
                                    AxesLabel->{r,z,V},
                                    Prolog->Thickness[0.001],
                                    ViewPoint->{1.3,-2.4,2}]]]


(* --- center of mass motion in the Penning trap --- *)

PenningCMPlot[x0_,y0_,x0d_,y0d_,w_]:= Block[{w0, a1, b1},
(* --- fix parameters Omega_0 = 1.0 --- *)
     w0 = 1.0;
     a1 = 0.25;
     b1 = 0.0;
     If[w <= 2*w0,Print[" "];
               Print[" cyclotron frequency too small"];
               Print[" choose w > 2"];
(* --- determine the amplitudes from the initial conditions --- *)
     gl1 = 2*ar + 2*br - x0 == 0;
     gl2 = -2*ai - 2*bi - y0 == 0;
     gl3 = 2*bi*w1 + 2*ai*w2 - x0d == 0;
     gl4 = 2*br*w1 + 2*ar*w2 - y0d == 0;
     result = Flatten[N[Solve[{gl1,gl2,gl3,gl4},{ar,ai,br,bi}]]];
(* --- solutions for the center of mass motion --- *)
x = 2*br*Cos[w1*t] + 2*bi*Sin[w1*t] + 2*ar*Cos[w2*t] + 2*ai*Sin[w2*t];
y = 2*ar*Sin[w2*t] - 2*ai*Cos[w2*t] + 2*br*Sin[w1*t] + 2*bi*Cos[w1*t];
z = a1*Cos[Sqrt[2 w0]*t + b1];
(* --- define frequencies --- *)
w1 = wc/2 + Sqrt[(wc/2)^2 - w0];
w2 = wc/2 - Sqrt[(wc/2)^2 - w0];
(* --- substitute the results result into the variables x, y, and z --- *)
x = Simplify[x /. result];
y = Simplify[y /. result];
x1 = x /. wc -> w;
x2 = y /. wc -> w;
x3 = z /. wc -> w;
(* --- plot the solution --- *)
ParametricPlot3D[{x1,x2,x3},{t,0,60},AxesLabel->{"x","y","z"},
```

```
                    PlotPoints->1000,
                    Prolog->Thickness[0.001]]
        ]]
  End[]
  EndPackage[]
```

3.4 Exercises

1. Create some pictures for a quadrupole arrangement of charges using the package **PointCharge'**. Choose the location of the charges in the representation plane of the potential section. What changes are required if your choice of coordinates for the charges is outside the representation plane? Perform some experiments with a larger number of charges.

2. Examine the electric potential of a disk segment under several boundary conditions using the package **BoundaryProblem'**; e.g. $\Phi_0 = \sin(\phi)$ or $\Phi_0 = \phi$. What changes occur in the potential if we change the angle α? Examine the influence of the upper summation index N on the accuracy of the solution.

3. Study the dynamic properties of two ions in a Penning trap for:

   a) a vanishing angular momentum ($\nu = 0$) and different frequency ratios λ. Which λ values result in chaotic motion and in a regular motion of the particles?

   b) find solutions for $\nu \neq 0$, $\lambda = 1$, and $\lambda = 2$.

   c) examine the parameter combination $\nu = 0$ and $\lambda = 1/2$.

4. Develop a *Mathematica* function to combine the relative and center of mass coordinates for a representation of motion in real space for the two ion problem of a Penning trap.

5. Reexamine the Green's function formalism and discuss the problem of a rectangular boundary with one side carrying a constant charge distribution. The three other sides are fixed to the ground potential.

6. Examine a collection of three particles in a Penning trap.

7. Discuss the motion of two particles in a Penning trap for $\nu \neq 0$ and λ arbitrary.

4

Quantum mechanics

4.1 The Schrödinger equation

The development of quantum mechanics as a field of study required an equation that would adequately describe experimentally observed quantum mechanical properties, such as the spectroscopic properties of atoms and molecules. In 1926, Schrödinger wrote down the equation of motion for a complex field in close analogy to the iconic equation of optics. Today, it is known as the Schrödinger equation. The Schrödinger equation for a single particle is

$$i\hbar\psi_t = -\frac{\hbar^2}{2m}\Delta\psi(\boldsymbol{x},t) + V(\boldsymbol{x})\psi(\boldsymbol{x},t), \qquad (4.1)$$

where $\psi(\boldsymbol{x},t)$ denotes the wave function, $V(\boldsymbol{x})$ an external potential representing the source of forces in the quantum system, $\hbar$ is Planck's constant, and m the mass of the particle under consideration.

The Schrödinger equation is a linear equation. It is well known that linear partial differential equations allow a superposition of their solutions to construct general solutions. Using this information with the two solutions ψ_1 and ψ_2 of the Schrödinger equation (4.1) allows us to construct the solution $\psi = c_1\psi_1 + c_2\psi_2$. We can identify Schrödinger's equation as a diffusion equation if we define an imaginary diffusion constant. To solve Schrödinger's equation, we can use the same solution procedure as for the diffusion equation. For certain initial values and known boundary values, we find the evolution of the wave function ψ by equation (4.1).

The main problem at the outset of quantum mechanics was the interpretation of the wave function ψ. Although Schrödinger's linear equation of motion (4.1) is completely deterministic, its solution $\psi(\boldsymbol{x},t)$ is not a measurable quantity. In fact, the only observable quantities in quantum mechanics are the probability $\psi^*\psi$ and any mean value based upon the distribution function ψ denoted by $\langle\psi|\Theta|\psi\rangle$.

Another consequence of the linearity of the Schrödinger equation is the property of dispersion. It is well known that linear equations of motion have dispersive waves as solutions. Since Schrödinger's equation (4.1) contains an imaginary factor i, we can expect the solutions for a free particle to undergo oscillations in the time domain. Plane waves are the simplest solutions to ψ. A particular solution of equation (4.1) with $V(x) = 0$ is given by

$$\psi_k(x,t) = \frac{1}{\sqrt{2\pi}} e^{i(kx-\omega(k)t)}. \tag{4.2}$$

The superposition of this particular solution delivers the general solution by

$$\psi(x,t) = \frac{1}{\sqrt{2\pi}} \int_{-\infty}^{\infty} A(k) e^{i(kx-\omega(k)t)} \, dk. \tag{4.3}$$

For simplicity's sake we limit our consideration to one spatial dimension. The solution (4.3) of the Schrödinger equation (4.1) is known as a wave packet. The spectral density $A(k)$ of the packet is completely determined by the initial condition $\psi(x, t = 0) = \psi_0(x)$. The representation (4.3) follows from the Fourier transform of the initial condition

$$A(k) = \frac{1}{\sqrt{2\pi}} \int_{-\infty}^{\infty} \psi_0(x) e^{-ikx} \, dx. \tag{4.4}$$

Inserting the spectral density into the general solution (4.3), we get the representation

$$
\begin{aligned}
\psi(x,t) &= \frac{1}{2\pi} \int_{-\infty}^{\infty} \int_{-\infty}^{\infty} \psi_0(x') e^{i(k(x-x')-\omega(k)t)} \, dk \, dx' \\
&= \int_{-\infty}^{\infty} dx' \, \psi_0(x') G(x, x', t),
\end{aligned}
\tag{4.5}
$$

where the Green's function is defined by

$$G(x, x', t) = \frac{1}{2\pi} \int_{-\infty}^{\infty} dk \, e^{i(k(x-x')-\omega(k)t)}. \tag{4.6}$$

The dispersion relation $\omega(k)$ of a dispersive wave is given by the defining equation of motion. For the Schrödinger equation with vanishing external potential $V(x) = 0$ the dispersion relation is $\omega(k) = \hbar k^2/(2m)$. Assuming a localized distribution $\psi_0(x) = \delta(x)$ for the initial condition of the wave function, we can write the related solution as follows

$$\psi(x,t) = \frac{1}{2\pi} \int_{-\infty}^{\infty} e^{ik(x-\alpha kt)} \, dk. \tag{4.7}$$

This initial condition (assumed to derive the wave function ψ) cannot be normalized. Although this assertion contradicts the quantum mechanical interpretation, our only interest here is to show the dispersive behavior of the wave function. The constant $\alpha = \hbar/(2m)$ is purely numerical. The relation (4.7) represents a solution of the Schrödinger equation (4.1) for the case of a free particle located at $x = 0$ with $t = 0$. Since the Schrödinger equation describes dispersive phenomena, we can observe a broadening of the wave packet diminishing for $t \to \infty$. Its shape is studied in the following. Replacing k by $k = \kappa/\sqrt{\alpha t}$ in equation (4.7), we obtain

$$\psi(x,t) = \frac{1}{\sqrt{\alpha t}} \frac{1}{2\pi} \int_{-\infty}^{\infty} \exp\left\{i(\kappa x/\sqrt{\alpha t} - \kappa^2)\right\} d\kappa. \tag{4.8}$$

Computing the square in the exponent, we get

$$\psi(x) = \frac{1}{2\pi} \frac{1}{\sqrt{\alpha t}} e^{ix^2/(4\alpha t)} \int_{-\infty}^{\infty} \exp\left\{-i(x/\sqrt{4\alpha t} - \kappa)^2\right\} d\kappa. \tag{4.9}$$

Substituting $\Gamma = x/(2\sqrt{\alpha t}) - \kappa$ gives us

$$\psi(x,t) = \frac{1}{2\pi\sqrt{\alpha t}} e^{ix^2/(4\alpha t)} \underbrace{\int_{-\infty}^{\infty} e^{i\Gamma^2} d\Gamma}_{= \sqrt{\pi} e^{i\pi/4}}$$

$$= \frac{1}{2\sqrt{\alpha t \pi}} \exp\left\{i\left(x^2/(4\alpha t) + \pi/4\right)\right\}. \tag{4.10}$$

This representation of the wave function for a free particle can be used to determine the probability of locating the particle at a certain time. As discussed above, ψ is not a function which is directly observable by experiment. To locate a particle at a certain location at a certain time, we have to study the probability distribution $|\psi|^2$ of the particle. The probability distribution of solution (4.10) is given by the expression

$$|\psi(x,t)|^2 = \frac{1}{4\alpha t \pi}. \tag{4.11}$$

This result shows that the probability of finding a free particle as described by Schrödinger's equation vanishes as time goes on. The probability of finding a particle at any location decreases with time and vanishes as $t \to \infty$. The dispersion process of the particle can be represented using *Mathematica* in a sequence of pictures. To animate the dispersion process we first define the wave function ψ of the free particle

```
Psi[x_,t_,hbar_:1,mass_:1] := Block[{alpha},
    alpha = hbar/(2 mass);
    Exp[I (x^2/(4 alpha t) + Pi/4)]/
    (2 Sqrt[alpha t Pi])
    ]
```

where mass m and $\hbar$ are set to unity. By an appropriate scaling of the coordinates we can eliminate these constants in the equation of motion. The probability distribution $|\psi|^2$ in relation (4.11) is only a function of time and does not show any spatial dependence. However, if we examine the wave function itself, we observe the spatial dispersion of the wave.

In figures 4.1 a time sequence of the real part of the wave function is plotted. The pictures are created by

```
figures =
Table[
  Plot[Evaluate[Re[Psi[x,t]]], {x,-6,6},
  AxesLabel->{"x","Re[Psi]"},
  PlotLabel->"t="<>ToString[t],
  Prolog->Thickness[0.001]],
  {t,0.5,3.0,0.5}];

figure = Show[GraphicsArray[Partition[figures,2]]]
```

The pictures show that the amplitude of the wave function decreases from about 0.5 to about 0.1 in a time range of 0.5 to 3.0. The dispersion of the wave packet is observable in the wave function. The wave function exhibits a reduced amplitude and a broadening of the initial packet.

The Schrödinger equation (4.1) not only describes time dependent properties of quantum mechanical systems but also stationary properties of these systems. Contrary to our observations about free particles, we now find that Schrödinger's equation describes stable particles. One central question for such a system is how to uncover its intrinsic characteristics such as the spectral properties. In the following, we examine one of the fundamental models of quantum mechanics—the harmonic oscillator.

Before discussing the spectral properties of the harmonic oscillator, we first summarize the solution steps for the time dependent Schrödinger equation by a short graphical representation given in figure 4.2.

   i) Starting point of the solution procedure is the partial differential equation (PDE) (4.1) and the initial solution of the wave function $\psi(x,0)$.

   ii) The use of the Fourier transform allows us to derive the spectral density $A(k)$ from the initial conditions.

  iii) A complete representation in Fourier space is attained when considering the time evolution which is given by the dispersion relation $\omega(k)$.

  iv) The inversion of the representation in Fourier space delivers the solution of the Schrödinger equation.

A similar solution procedure for nonlinear PDEs is discussed in Chapter 5.

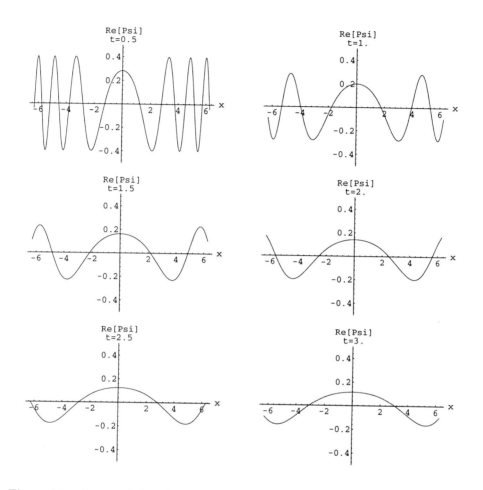

Figure 4.1. Time evolution of a wave packet for the Schrödinger equation. Initial conditions are $\psi_0(x) = \delta(x)$.

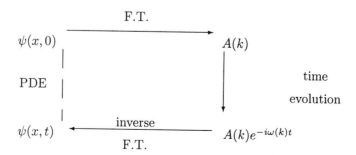

Figure 4.2. Solution steps for a linear PDE by using the Fourier transform.

4.2   One dimensional potential

In quantum mechanics, the measurement of a physical quantity A can result only in one of the eigenvalues of the corresponding operator $\hat{A}$. The eigenvalues of $\hat{A}$ forming the spectrum of the operator may be discrete, continuous, or both. The eigenfunctions of $\hat{A}$ form a complete basis which may be used to expand an arbitrary wave function. The expansion coefficients can be used to determine the probability of finding the system in an eigenstate of the operator A with eigenvalue a. Central to quantum mechanics is the determination of these eigenvalues and their related eigenfunctions.

One of the fundamental quantities of a quantum dynamical system is its energy. The operator corresponding to energy is the Hamiltonian operator of the system. The Hamiltonian for a particle with mass m located in a potential V is represented by $\hat{H} = -\hbar^2/(2m)\Delta + V(\boldsymbol{x})$. The determination of eigenvalues and eigenfunctions is demonstrated with a one dimensional model, the potential well. The potential well of depth $V = -V_0$ discussed in the following extends between $-a \leq x \leq a$ where a is the maximum extension. Beyond the maximum extension the potential vanishes. A graphical representation of the potential well of depth V is given in figure 4.3.

We study the case for which the kinetic energy of the particle is smaller than the minimum potential value V_0; i.e., $T < V_0$. The total energy E of the system is $E = T - V_0 < 0$. The particle has a negative total energy in the domains 1 and 3 depicted in figure 4.3. In classical mechanics the particle cannot be found in these regions. Contrary to classical mechanics, however, quantum mechanics allows the existence of particles in regions where they are classically forbidden. The domains 1 and 3 are governed by the eigenvalue equations $\hat{H}\psi = E\psi$ which are given in the differential representation by

$$\psi'' - \kappa^2\psi = 0 \tag{4.12}$$

where $\kappa^2 = -2mE/\hbar^2 > 0$ is a positive constant containing the total energy. Primes denote differentiation with respect to the spatial coordinate. The solution of equation (4.12) represents the domains 1 and 3 by

$$
\begin{aligned}
\psi_1 &= A_1 e^{\kappa x} + B_1 e^{-\kappa x} \quad \text{for} \quad -\infty < x \leq -a & (4.13)\\
\psi_3 &= A_3 e^{\kappa x} + B_3 e^{-\kappa x} \quad \text{for} \quad a \leq x < \infty. & (4.14)
\end{aligned}
$$

In domain 2; the eigenvalue equation takes the form

$$\psi'' + k^2\psi = 0 \tag{4.15}$$

where $k^2 = 2m(V_0 + E)/\hbar^2 > 0$. The complete solution of (4.15) is given by

$$\psi_2 = A_2 \cos(kx) + B_2 \sin(kx) \quad \text{for} \quad -a \leq x \leq a. \tag{4.16}$$

From the normalization condition it follows that the eigenfunctions given by relations (4.13) and (4.14) require that the coefficients B_1 and A_3 vanish; i.e.,

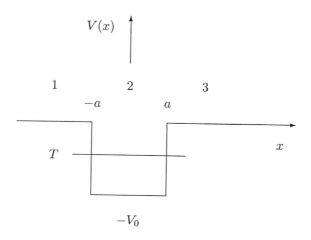

Figure 4.3. The potential well of depth V.

$B_1 = A_3 = 0$. The remaining parameters A_1, B_2, A_2, and B_3 are determined by applying the continuity condition of the wave function and its first derivative at the endpoints of the potential well $(x = -a$ and $x = a)$.

$$\psi_1 = \psi_2 \quad \text{and} \quad \psi_1' = \psi_2' \quad \text{for} \quad x = -a \qquad (4.17)$$
$$\psi_2 = \psi_3 \quad \text{and} \quad \psi_2' = \psi_3' \quad \text{for} \quad x = a. \qquad (4.18)$$

The four equations form a homogeneous system of equations for the unknowns A_1, B_3, A_2, and B_2. In a matrix representation we get

$$\begin{pmatrix} e^{-\kappa a} & -\cos(ka) & \sin(ka) & 0 \\ \kappa e^{-\kappa a} & -k\sin(ka) & k\cos(ka) & 0 \\ 0 & -\cos(ka) & -\sin(ka) & e^{-\kappa a} \\ 0 & k\sin(ka) & -k\cos(ka) & -\kappa e^{-\kappa a} \end{pmatrix} \begin{pmatrix} A_1 \\ A_2 \\ B_2 \\ B_3 \end{pmatrix} = 0. \qquad (4.19)$$

A nontrivial solution of equation (4.19) exists if the determinant of the matrix vanishes. This condition delivers the relation

$$\kappa^2 - k^2 + 2\kappa k \cot(2ka) = 0 \qquad (4.20)$$

with solutions

$$\kappa = k\tan(ka) \qquad (4.21)$$
$$\kappa = -k\cot(ka). \qquad (4.22)$$

If we consider the first of these relations (4.21), we find that $B_2 = 0$, $B_3 = A_1$, and $A_2\cos(ka) = A_1 e^{-\kappa a}$. The second relation, (4.22), results in the conditions

$A_2 = 0$, $B_3 = -A_1$ and $B_2 \sin(ka) = -A_1 e^{-\kappa a}$. We can thus distinguish between two systems of eigenfunctions: a symmetric one and an antisymmetric one. The symmetry of the eigenfunctions is obvious if we exchange the coordinates by $x \to -x$. The symmetrical case is represented by

$$\kappa = k \tan(ka) \tag{4.23}$$

$$\psi_1 = A_1 e^{\kappa x} \tag{4.24}$$

$$\psi_2 = A_1 e^{-\kappa x} \frac{\cos(kx)}{\cos(ka)} \tag{4.25}$$

$$\psi_3 = A_1 e^{-\kappa x}. \tag{4.26}$$

The antisymmetric case follows from the relations

$$\kappa = -k \cot(ka) \tag{4.27}$$

$$\psi_1 = -A_1 e^{\kappa x} \tag{4.28}$$

$$\psi_2 = A_1 e^{-\kappa x} \frac{\sin(kx)}{\sin(ka)} \tag{4.29}$$

$$\psi_3 = A_1 e^{-\kappa x}. \tag{4.30}$$

From the normalization condition

$$\int_{-\infty}^{\infty} dx\, \psi^2 = \int_{-\infty}^{-a} dx\, \psi_1^2 + \int_{-a}^{a} dx\, \psi_2^2 + \int_{a}^{\infty} dx\, \psi_3^2 = 1 \tag{4.31}$$

we get a relation for the undetermined amplitude A_1

$$\frac{1}{A_1^2} = ae^{-2\kappa a} \left(1 + \frac{1}{\kappa a} + \frac{\kappa}{k^2 a} + \frac{\kappa^2}{k^2} \right). \tag{4.32}$$

Relation (4.32) is satisfied for both the symmetric and antisymmetric eigenfunctions. To calculate the eigenvalues, note that $\kappa^2 + k^2 = 2mV_0/\hbar^2 > 0$ is independent of the total energy E. If we introduce the parameter

$$C^2 = a^2 \frac{2mV_0}{\hbar^2} = (\kappa^2 + k^2)a^2, \tag{4.33}$$

we can eliminate κ from the eigenvalue equations. The equations determining the eigenvalues are now

$$\frac{\sqrt{C^2 - (ka)^2}}{ka} = \tan(ka) \tag{4.34}$$

$$-\frac{ka}{\sqrt{C^2 - (ka)^2}} = \tan(ka). \tag{4.35}$$

Using relation (4.34) or (4.35), we can calculate ka and $E = \hbar^2 k^2 - 2mV_0$.

The problem with the potential well is not the derivation of its solution but the calculation of the eigenvalues from equations (4.34) and (4.35). We use *Mathematica* to solve the problem numerically for varying well depths V_0 and well widths a. The left hand and right hand sides of equations (4.34) and (4.35) are graphically represented in figure 4.4 for $V_0 = 12$ and $a = 1$.

Figure 4.4 is created by means of the function **Spectrum[12,1]** defined in the package **QuantumWell'**. Also defined in the package **QuantumWell'** are the eigenfunctions **PsiSym[]** and **PsiASym[]**. The function **Spectrum[]** provides us with a graphical representation of the eigenfunctions and prints out the related eigenvalues in a list. Some examples of these eigenfunctions are given in figures 4.5 and 4.6. The function **Spectrum[]** creates a sequence of eigenfunction pictures starting with the symmetric ones and followed by the antisymmetric ones. Figures 4.5 and 4.6 contain the superposition of these sequences into one picture.

```
———————————————— File quantumw.m ————————————————

BeginPackage["QuantumWell'"]

Clear[PsiSym,PsiASym,Spectrum]

PsiSym::usage= "PsiSym[x_,k_,a_] determines the symmetric eigenfunction for
a potential well of depth -VO. The input parameter k fixes the energy and
2a the width of the well. PsiSym is useful for a numerical representation
of eigenfunctions."

PsiASym::usage= "PsiASym[x_,k_,a_] determines the antisymmetric eigenfunction
for a potential well of depth -VO. The input parameter k fixes the energy and
2a the width of the well. PsiASym is useful for a numerical representation
of eigenfunctions."

Spectrum::usage= "Spectrum[VO_,a_] calculates the negative eigenvalues in a
potential well. VO is the potential depth and 2a the width of the well. The
eigenvalues are returend as a list and are available in the variables lsym
and lasym as replacement rules. The corresponding plots of eigenfunctions
are stored in the variables Plsym and Plasym. The determining equation for
the eigenvalues is plotted."

(* --- define global variables --- *)

Plsym::usage="Variables containing the symmetric plots of the
eigenfunctions."
Plasym::usage="Variables containing the antisymmetric plots of the
eigenfunctions."
lsym::usage="List of symmetric eigenvalues."
lasym::usage="List of antisymmetric eigenvalues."
k:usage="Eigenvalue."

Begin["'Private'"]

(* --- symmetric eigenfunctions --- *)

PsiSym[x_,k_,a_]:=Block[{kapa, A1},
     kapa = k Tan[k a];
(* --- normalization constant --- *)
     A1 = 1/Sqrt[a Exp[-2 a kapa] (1+1/(kapa a)+kapa/(k^2 a) +
          kapa^2/k^2)];
(* --- define the three domains of solution --- *)
```

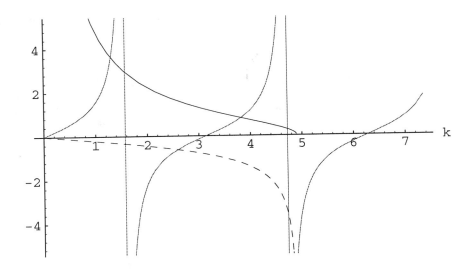

Figure 4.4. Graphical representation of the eigenvalue equation for $V_0 = 12$ and $a = 1$. The solid curves represent the symmetrical case while the dashed curves represent the antisymmetric case. The right hand side of the eigenvalue equation reads $\tan(ka\#)$.

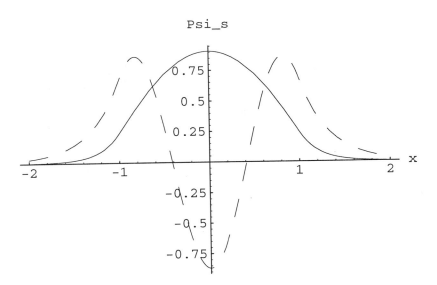

Figure 4.5. The symmetric eigenfunctions for a potential well with depth $V_0 = 12$ and width $a = 1$. For the given potential depth, there are a total of four eigenvalues, two of which are shown in this figure while the other two are shown in the next figure. The solid eigenfunction with a broad single maximum and no nodes is related to the lowest eigenvalue $k = 1.30183$ of the symmetric case. The second symmetric eigenvalue is $k = 3.81858$. The corresponding eigenfunction is dashed.

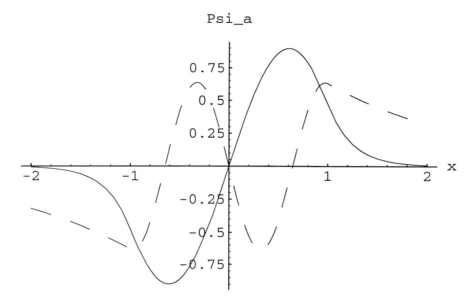

Figure 4.6.   The antisymmetric eigenfunction for the potential with $V_0 = 12$ and $a = 1$. The
two antisymmetric eigenfunctions are correlated with the eigenvalues $k = 2.5856$ and
$k = 4.85759$. The first eigenfunction is represented by a solid curve and the second is dashed.

```
        Which[-Infinity < x && x < -a, A1 Exp[kapa x],
              -a <= x && x <= a, A1 Exp[-kapa a] Cos[k x]/Cos[k a],
              a < x && x < Infinity, A1 Exp[-kapa x]
              ]
        ]

(* --- antisymmetric eigenfunctions --- *)

PsiASym[x_,k_,a_]:=Block[{kapa, A1},
        kapa = -k Cot[k a];
(* --- normalization constant --- *)
        A1 = 1/Sqrt[a Exp[-2 a kapa] (1+1/(kapa a)+kapa/(k^2 a) +
             kapa^2/k^2)];
(* --- define the three domains of solution --- *)
        Which[-Infinity < x && x < -a, -A1 Exp[kapa x],
              -a <= x && x <= a, A1 Exp[-kapa a] Sin[k x]/Sin[k a],
              a < x && x < Infinity, A1 Exp[-kapa x]
              ]
        ]

(*--- determination of the eigenvalues; plot of the eigenfunctions ---*)

Spectrum[V0_,a_]:=Block[
       {hbar=1, m=1, ymax, C2, rhs, lhssym, lhsasym, equatsym,
        equatasym, kmax, nsym, nasym, resultsym, resultasym},
(* --- define constants and the eigenvalue equation --- *)
```

```
        C2 = 2 m V0 a^2/(hbar^2);
        rhs = Tan[k a];
        lhssym = Sqrt[C2-(k a)^2]/(k a);
        lhsasym = -k a/Sqrt[C2-(k a)^2];
        equatsym = Sqrt[C2-(k a)^2]/(k a) - Tan[k a];
        equatasym = -k a/Sqrt[C2-(k a)^2]-Tan[k a];
(* --- location of the singularity in k --- *)
        kmax = Sqrt[C2/a^2];
(* --- number of symmetric eigenvalues --- *)
        nsym = Floor[N[kmax/(Pi/a)]] + 1;
(* --- number of antisymmetric eigenvalues --- *)
        nasym = Floor[N[(kmax - Pi/(2 a))/(Pi/a)]] + 1;
(* --- initialize the lists for the eigenvalues --- *)
        lsym = {};
        lasym = {};
(* --- calculate the symetric eigenvalues --- *)
        Do[
        resultsym = Chop[FindRoot[equatsym==0,{k,0.1+(Pi/a) (i-1)}]];
        AppendTo[lsym,resultsym],
        {i,1,nsym}];
     (* --- Chop[] replaces small numbers (<10^(-10)) by 0 --- *)
(* --- calculate the antisymmetric eigenvalues --- *)
        Do[
        resultasym = Chop[FindRoot[equatasym==0,{k,Pi/(2 a)+
                                     0.1+(Pi/a) (i-1)}]];
        AppendTo[lasym,resultasym],
        {i,1,nasym}];
(* --- plot the eigenvalue equation --- *)
        ymax = lhssym 1.5 /. lsym[[1]];
    Off[Plot::plnr];
    Plot[{rhs,lhssym,lhsasym},{k,0.01,3 kmax/2},
                 PlotRange->{-ymax,ymax},
                 Prolog->Thickness[0.001],
                 PlotStyle->{RGBColor[1,0,0],Dashing[{}],Dashing[{1/60}]},
                 AxesLabel->{"k"," "}];
    On[Plot::plnr];
(* --- plot the symmetric eigenfunctions --- *)
        Do[
        k1 = k /. lsym[[i]];
        Plsym[i] = Plot[PsiSym[x,k1,a],{x,-2 a,2 a},
                     AxesLabel->{"x","Psi_s"},
                     PlotRange->All,
                     Prolog->Thickness[0.001],
                     PlotStyle->{Dashing[{1/(i 20)}]}],
        {i,1,nsym}];
(* --- plot the antisymmetric eigenfunctions --- *)
        Do[
        k1 = k /. lasym[[i]];
        Plasym[i] = Plot[PsiASym[x,k1,a],{x,-2 a,2 a},
                     AxesLabel->{"x","Psi_a"},
                     PlotRange->All,
                     Prolog->Thickness[0.001],
                     PlotStyle->{Dashing[{1/(i 20)}]}],
        {i,1,nasym}];
(* --- print the eigenvalues --- *)
        Print[" "];
        Print[" ----  eigenvalues ---- "];
```

```
Print[" "];
Do[
k1 = k /. lsym[[i]];
If[i<=nasym,
   k2 = k /. lasym[[i]],
   k2 = "---"];
Print[" sym eigenvalue k",i," = ",k1," asym eigenvalue k",
         i," = ",k2],
{i,1,nsym}]
]
End[]
EndPackage[]
```

4.3 The harmonic oscillator

The potential energy for a stable system exhibits a local minimum. One of the standard methods of physics is to expand the potential energy around the point of a local minimum in a Taylor series,

$$V = V(0) + \frac{1}{2} \left(\frac{\partial^2 V}{\partial x^2} \right) \bigg|_{x=0} x^2 + \dots \qquad (4.36)$$

where x denotes the displacement from the equilibrium point. The potential satisfies $\partial V/\partial x = 0$ at the stable equilibrium point. If the particle of mass m only undergoes small oscillations around the equilibrium point the first two terms of relation (4.36) are sufficient to describe the potential energy. Choosing the origin of the energy to be identical with $V(0)$ of the expansion, we can express the Hamiltonian of the harmonic oscillator

$$H_{cl} = \frac{p^2}{2m} + \frac{k}{2} x^2 \qquad (4.37)$$

where $k = \frac{\partial^2 V}{\partial x^2}\big|_{x=0}$ is the spring constant of the oscillator. We already know that the classical solution for the harmonic oscillator is given by a periodic function

$$x(t) = A\cos(\omega t + \beta) \qquad \text{where} \qquad \omega = \sqrt{\frac{k}{m}} \qquad (4.38)$$

and the system undergoes harmonic oscillations around the equilibrium point. The time average of the total energy follows from relations (4.37) and (4.38)

$$E = \frac{1}{2}mA^2\omega^2 = m\omega^2\bar{x^2}; \tag{4.39}$$

i.e., the time–averaged energy depends quadratically on the amplitude A of the oscillations.

In this section our aim is to examine the quantum mechanical properties of the harmonic oscillator and compare them with the classical situation. The transition from classical to quantum mechanics is formally achieved by replacing the classical coordinates with quantum mechanical operators: $x \to \hat{x} = x$, and $p \to \hat{p} = -i\hbar\partial_x$. Using the transformations in the Hamiltonian yields the timeless Schrödinger equation in the form of an eigenvalue problem given by

$$\left\{ \frac{d^2}{dx^2} - \frac{\omega^2 m^2 x^2}{\hbar^2} + \frac{2mE}{\hbar^2} \right\} \psi(x) = 0 \tag{4.40}$$

where ψ denotes the eigenfunctions of the Hamiltonian. By an appropriate scaling of the spatial coordinate $\xi = x\sqrt{m\omega/\hbar}$ and of the eigenvalue $\epsilon = 2E/(\hbar\omega)$, we get the eigenvalue problem in a standard form

$$\left(\frac{d^2}{d\xi^2} - \xi^2 + \epsilon \right) \psi(\xi) = 0. \tag{4.41}$$

The question here is what type of function $\psi(\xi)$ satisfies equation (4.41). As a solution we try the expression

$$\psi(\xi) = v(\xi)e^{-\xi^2/2}. \tag{4.42}$$

From equation (4.41) it follows that the amplitude v has to satisfy the ODE

$$v'' - 2\xi v' + (\epsilon - 1)v(\xi) = 0 \tag{4.43}$$

where primes denote differentiation with respect to ξ. To be physically acceptable the wave function $\psi(\xi)$ must be continuous and finite. The amplitude $v(\xi)$ defined by equation (4.43) is a finite function if v is a polynomial of finite order. This type of solutions exists if

$$\epsilon = 2n + 1 \qquad \text{where} \qquad n = 0, 1, 2, \ldots. \tag{4.44}$$

For each value n there exists a polynomial of order n which satisfies equation (4.43). These polynomials are known as Hermite polynomials, defined by

$$H_n(\xi) = (-1)^n e^{\xi^2} \frac{d^n}{d\xi^n} e^{-\xi^2}. \tag{4.45}$$

In *Mathematica*, the Hermite polynomials are identified by the function **HermiteH[ ]**. The wave function ψ of the harmonic oscillator is represented in scaled coordinates by

$$\psi_n(\xi) = \frac{1}{\sqrt{n!2^n\sqrt{\pi}}} H_n(\xi) e^{-\xi^2/2}. \tag{4.46}$$

The corresponding eigenvalues of the harmonic oscillator are

$$E_n = \hbar\omega\left(n + \frac{1}{2}\right). \tag{4.47}$$

Each eigenvalue has its own eigenfunction which is either even or odd with respect to coordinate reflections in ξ. Note that the eigenvalues and eigenfunctions have a one-to-one correspondence (i.e., the spectrum is non–degenerated). The first four even and odd eigenfunctions of the harmonic oscillator are depicted in figures 4.7 and 4.8.

The probability distribution $|\psi|^2$ of finding the harmonic oscillator in a certain state n in the range $\xi \pm d\xi$ is given by

$$|\psi|^2 d\xi = \frac{1}{n!2^n\sqrt{\pi}} H_n^2(\xi) e^{-\xi^2} d\xi = w_{qm}(x)d\xi. \tag{4.48}$$

The classical probability of finding a particle in the range $x \pm dx$ is determined by the period T of the oscillator.

$$w_{cl}(x) = \frac{dt}{T} = \frac{\omega}{2\pi} \frac{dx}{|v|} \tag{4.49}$$

where $x(t)$ is represented by the classical solution (4.38). The corresponding velocity v follows from the time derivative of x

$$v = -A\omega\sqrt{1 - \left(\frac{x}{A}\right)^2}. \tag{4.50}$$

In scaled variables ξ we find for the classical probability the relation

$$w_{cl}(\xi) = \frac{1}{2\pi\sqrt{2n+1}} \frac{1}{\sqrt{1 - \dfrac{\xi^2}{2n+1}}}. \tag{4.51}$$

Specifying either the energy or the eigenvalue of the harmonic oscillator enables us to compare the classical probability with the quantum mechanical result. A graphical representation of these two quantities is given in figures 4.9 and 4.10. Figure 4.9 shows the ground state and figure 4.10 shows the eigenvalue with $n = 5$. It can be clearly seen that the quantum mechanical behavior of the probability density is different from its classical behavior. In the classical case, the particle spends most of its time near the two turning points where the density $|\psi|^2$ is large. Quantum mechanically there is a high probability that the particle is located near

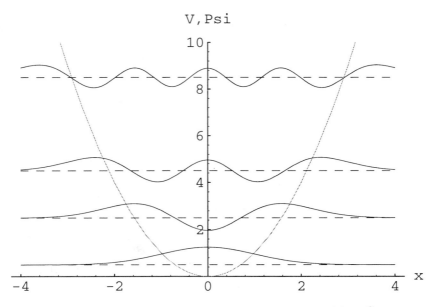

Figure 4.7. Symmetric eigenfunctions of the harmonic oscillator $V(x) = x^2$ for eigenvalues $n = 0, 2, 4, 8$. The eigenfunctions are centered around the energetic levels $E/(\hbar\omega) = n + 1/2$ corresponding to the eigenvalues n.

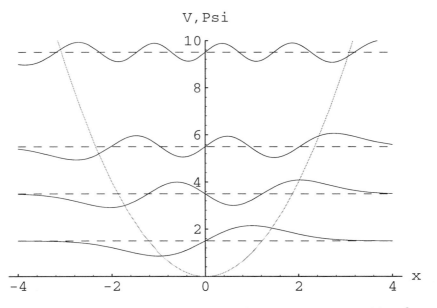

Figure 4.8. Antisymmetric eigenfunctions of the harmonic oscillator $V(x) = x^2$ for eigenvalues $n = 1, 3, 5, 9$. The eigenfunctions are centered around the energy levels $E/(\hbar\omega) = n + 1/2$ corresponding to the eigenvalues n.

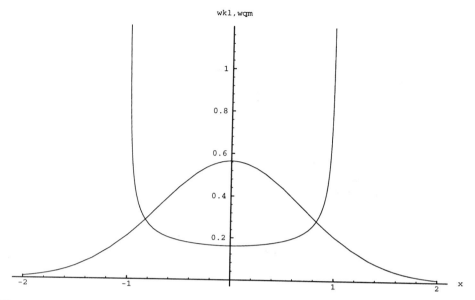

Figure 4.9. Classical and quantum mechanical probability density for the harmonic oscillator in the ground state. The classical probability shows a singular behavior at the turning points of the motion.

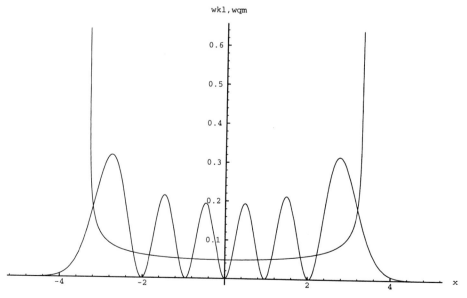

Figure 4.10. Comparison between the classical and quantum mechanical probability density for the eigenvalue $n = 5$. The singular points of the classical probability w_{cl} are located at $x = \pm 3.316$.

the center of the potential (ground state). In an excited state, we observe regions where the particle cannot be found (see figure 4.10). This is due to the fact that the quantum mechanical probability density oscillates for $n > 0$, which in turn is a consequence of the oscillations of the wave function.

At the classical turning points a completely different behavior of the quantum particle is apparent. Where the classical particle cannot be found in quantum mechanics, there is a finite probability for locating a particle outside the potential well. This tunneling of the particle into the potential barrier is unusual and cannot be explained by classical mechanics.

The eigenfunctions and the harmonic potential $V(\xi)$ are superimposed on each other in figures 4.7 and 4.8. The related classical and quantum mechanical probabilities are shown in figures 4.9 and 4.10. The functions to create these figures for certain eigenvalues are contained in the package **HarmonicOscillator'**.

The given derivation of the wave function is based on the defining equation of the Hermite polynomials (4.41). The solution of the scaled equation (4.41) delivers the complete set of eigenfunctions in one step. In the following we show how the set of eigenfunctions can be derived by an iterative procedure involving creation and annihilation operators. All the eigenfunctions are created out of the ground state of the harmonic oscillator,

$$\psi_0(x) = \frac{1}{\pi^{1/4}} e^{-\xi^2/2}. \tag{4.52}$$

The whole set of eigenfunctions can be created using the following "creation" a^+ and "annihilation" a operators

$$a^+ = \frac{1}{\sqrt{2}} \left(\xi - \frac{\partial}{\partial \xi} \right) = \frac{1}{\sqrt{2}} \left(\hat{\xi} - i\hat{p} \right) \tag{4.53}$$

$$a = \frac{1}{\sqrt{2}} \left(\xi + \frac{\partial}{\partial \xi} \right) = \frac{1}{\sqrt{2}} \left(\hat{\xi} + i\hat{p} \right) \tag{4.54}$$

The name of the operators stems from the action of the wave functions respectively creating and annihilating a quantum mechanical state. The actions of operators a^+ and a can be demonstrated by introducing two *Mathematica* functions **a[ ]** and **across[ ]**. The definitions are given below and use the representations given in equations (4.53) and (4.54).

```
a[psi_, xi_:x]  := (xi*psi + D[psi, xi])/Sqrt[2]

across[psi_, xi_:x]  := (xi*psi - D[psi, xi])/Sqrt[2]
```

If we apply the defined functions to the ground state we get the first excited state or simply zero. The definition of the ground state is contained in the function **Psi[ ]**. We get

```
Psi[xi_,n_]:= HermiteH[n,xi] Exp[-xi^2/2]/Sqrt[n! 2^n Sqrt[Pi]]

across[Psi[x,0]];


  Sqrt[2] x
  -----------
    2
  x /2    1/4
E       Pi


a[Psi[x,0]]


0
```

Comparing the *Mathematica* result with the first excited state ψ_1, we find that they are equivalent.

```
across[Psi[x,0]] == Psi[x,1] // Simplify

True
```

The higher eigenfunctions are derived from the ground state by the relation

$$\psi_n(\xi) = \frac{1}{\sqrt{n!}}(a^+)^n\psi_0(\xi). \qquad (4.55)$$

In *Mathematica* repeatedly applying an operator is achieved using the function **Nest[ ]**.

```
Nest[across,Psi[x,0],5] //Simplify

                   2      4
Sqrt[2] x (15 - 20 x  + 4 x )
-----------------------------
            2
          x /2   1/4
        E      Pi
```

We assume that **Psi[ ]** is a function of x. When using **Nest[ ]**, we can repeatedly apply the function **across[ ]** to the wave function **Psi[ ]**. The number of applications of **across[ ]** to **Psi[ ]** is controlled by the second argument of **Nest[ ]**. In the above example, we applied **across[ ]** five times to **Psi[ ]**. The result is the representation of ψ_5. If we are interested in the functions preceding ψ_5, we can use **NestList[ ]** instead.

```
PsiList = NestList[across,Psi[x,0],5] //Simplify

                                   2
         1            Sqrt[2] x    -1 + 2 x
{-----------, ------------, -----------,
     2               2              2
   x /2   1/4     x /2   1/4     x /2   1/4
  E      Pi      E      Pi      E      Pi

                           2          2       4
   Sqrt[2] x (-3 + 2 x )  3 - 12 x  + 4 x
   --------------------, ----------------,
           2                     2
         x /2   1/4            x /2   1/4
        E      Pi             E      Pi

                        2       4
   Sqrt[2] x (15 - 20 x  + 4 x )
   ----------------------------}
             2
           x /2   1/4
          E      Pi
```

The unnormalized wave functions contained in the list **PsiList** are eigenfunctions of the harmonic oscillator. To determine the normalization factors, we integrate **PsiList** over the total space

```
norm = 1/Sqrt[Integrate[Expand[PsiList^2],{x,-Infinity,Infinity}]]

           1        1         1          1
{1, 1, -------, -------, ---------, ----------}
       Sqrt[2]  Sqrt[6]  2 Sqrt[6]  2 Sqrt[30]
```

The normalized eigenfunctions are now given by

```
PsiList = PsiList norm

                                               2
         1            Sqrt[2] x      -1 + 2 x
{-----------, ------------, --------------------,
     2               2               2
   x /2   1/4     x /2   1/4          x /2   1/4
  E      Pi      E      Pi     Sqrt[2] E      Pi

                     2          2       4
       x (-3 + 2 x )          3 - 12 x  + 4 x
      -------------------, --------------------,
               2                     2
```

```
           x /2   1/4                   x /2   1/4
  Sqrt[3] E      Pi     2 Sqrt[6] E      Pi


                 2       4
  x (15 - 20 x  + 4 x  )
  ---------------------}
                 2
                 x /2   1/4
  2 Sqrt[15] E      Pi
```

The preceding functions are collected in the package **HarmonicOscillator'** which is listed below.

```
──────────────────── File harmonic.m ────────────────────

BeginPackage["HarmonicOscillator'"]

Clear[a,across,Psi,wcl,wqm]

Psi::usage = "Psi[xi_,n_] represents the eigenfunction of the harmonic
oscillator. The first argument xi is the spatial coordinate. The second
argument n fixes the eigenstate."

wcl::usage = "wcl[xi_,n_] calculates the classical probability of locating
the particle in the harmonic potential. The first argument xi is the spatial
coordinate while n determines the energy given as eigenvalue."

wqm::usage = "wqm[xi_,n_] calculates the quantum mechanical probability for
an eigenvalue state n. The first argument represents the spatial
coordinate."

a::usage = "a[psi_, xi_:x] annihilation operator for eigenfunction psi. The
second argument specifies the independent variable of the function psi."

across::usage = "across[psi_, xi_:x] creation operator for eigenfunction
psi. The second argument specifies the independent variable of psi."

x::usage

Begin["'Private'"]

(* --- eigenfunctions of the harmonic oscillator --- *)

Psi[xi_,n_]:= HermiteH[n,xi] Exp[-xi^2/2]/Sqrt[n! 2^n Sqrt[Pi]]

(* --- classical probability distribution of the harmonic oscillator ---*)

wcl[xi_,n_]:= 1/(Sqrt[2 n+1] Sqrt[1-(xi/Sqrt[2 n+1])^2] 2 Pi)

(* --- quantummechanical probability distribution of the harmonic oscillator
       --- *)

wqm[xi_,n_]:= Psi[xi,n]^2

(* --- annihilation operator --- *)
```

```
a[psi_, xi_:x] := (xi psi + D[psi, xi])/Sqrt[2]

(* --- creation operator --- *)

across[psi_, xi_:x] := (xi psi - D[psi, xi])/Sqrt[2]

End[]
EndPackage[]
```

4.4 Anharmonic oscillator

So far we have discussed problems which assume harmonic particle motion. In real systems, harmonic motion is the exception rather than the rule. In general, forces are not proportional to linear displacements. From the example of the pendulum in classical mechanics (see section 2.3), we recall that the restoring force is not proportional to linear displacements. Another example is that of large molecules in quantum chemistry: in contrast to the binding potential of a diatomic molecule [4.2], the forces between atoms in a large molecule are anharmonic.

The classical work on anharmonic forces in quantum mechanics was done by Pöschel & Teller [4.3] who examined the single anharmonic oscillator. Lotmar [4.4] in 1935 studied an ensemble of anharmonic oscillators and established their connection with large molecules. We examine here an altered Pöschel–Teller potential which today is used in the inverse scattering method of solving nonlinear evolution equations (see chapter 5). The interaction potential for a quantum mechanical system was given by Flügge [4.5] in the form

$$V(x) = -V_0 \operatorname{sech}^2(x) \tag{4.56}$$

where V_0 is a constant determining the depth of the potential well. The related stationary Schrödinger equation in scaled variables reads

$$\psi_{xx} + (\lambda + V_0 \operatorname{sech}^2(x))\psi = 0. \tag{4.57}$$

In our examination we determine the eigenvalues $\lambda = 2mE/\hbar^2$ which depend on the potential depth V_0. Another point of our study is the form of the wave functions in the asymptotic range $|x| \to \infty$. We first introduce some changes in the notation of equation (4.57). Substituting for the independent variable x using the relation $\xi = \tanh(x)$ in equation (4.57), we get the expression

$$(1 - \xi^2)\frac{d}{d\xi}\left\{(1 - \xi^2)\frac{d\psi}{d\xi}\right\} + \left\{\lambda V_0(1 - \xi^2)\right\}\psi = 0 \quad \text{where} \quad -1 < \xi < 1 \tag{4.58}$$

or the equivalent standard representation of equation (4.58)

$$\frac{d}{d\xi}\left\{(1-\xi^2)\frac{d\psi}{d\xi}\right\} + \left(V_0 + \frac{\lambda}{1-\xi^2}\right)\psi = 0. \tag{4.59}$$

Equation (4.59) is the defining equation for the associated Legendre polynomials. For the solution of equation (4.59) we assume that the potential depth is given by positive integer $V_0 = N(N+1)$ where N is a positive number. Equation (4.59) possesses discrete bound solutions in the range $\xi \in [-1,1]$ if and only if $\lambda = -n^2 < 0$ with $n = 1, 2, \ldots, N$. The eigenfunctions of the Schrödinger equation (4.59) are proportional to the associated Legendre functions $P_N^n(\xi)$ defined by

$$P_N^n(\xi) = (-1)^n (1-\xi^2)^{n/2} \frac{d^n}{d\xi^n} P_N(\xi), \tag{4.60}$$

where $P_N(\xi)$ are the Legendre polynomials of degree N

$$P_N(\xi) = \frac{1}{N!2^N} \frac{d^N}{d\xi^N}(\xi^2 - 1)^N. \tag{4.61}$$

The constant connecting the Legendre functions with the eigenfunctions of the Pöschel–Teller problem is a product of the normalization condition and the eigenfunctions. The following *Mathematica* function represents the eigenfunctions of the Pöschel–Teller system. The associated Legendre polynomials are given by the function **LegendreP[ ]**.

```
PoeschelTeller[x_, n_Integer, N_Integer] :=
  Block[{norm, integrand, xi},
    If[n <= N && n > 0,
  (* --- the associated Legendre polynomial specify the eigenfunction --- *)
      integrand = LegendreP[N, n, xi];
  (* --- determine the normalization constant --- *)
      norm = Integrate[integrand^2/(1-xi^2), {xi, -1, 1}];
  (* --- normalize the eigenfunctions --- *)
      integrand = integrand/Sqrt[norm] /. xi -> Tanh[x];
      Simplify[integrand],
  (* --- check errors in the input parameters --- *)
      If[N<n,
      Print["---  wrong argument n > N"]];
      If[n<0,
      Print["---  wrong argument n < 0"]]]
    ]
```

The eigenfunctions for $N = 4$ are

```
PoeschelTeller[x,1,4]

             2                        3
Sqrt[5] Sqrt[Sech[x] ] (3 Tanh[x] - 7 Tanh[x] )
-----------------------------------------------
                    4

PoeschelTeller[x,2,4]

                                  4
Sqrt[5] (-4 + 3 Cosh[2 x]) Sech[x]
----------------------------------
                4

PoeschelTeller[x,3,4]

               2 3/2
-(Sqrt[105] (Sech[x] )    Tanh[x])
---------------------------------
             4

PoeschelTeller[x,4,4]

     35       4
Sqrt[--] Sech[x]
     32
```

The results for $n = 1$ and $n = 3$ are graphically represented in figure 4.11:

```
Plot[{PoeschelTeller[x,1,4],PoeschelTeller[x,3,4]},
    {x,-5,5},PlotPoints->50,AxesLabel->{"x","Psi"}]
```

Using the option **PlotPoints $->$ 50** the number of plot points has been increased in order to get a smooth curve for the eigenfunctions.

So far we derived the discrete spectrum of the modified Pöschel–Teller problem. In the following we consider the continuous eigenvalues $\lambda = k^2 > 0$ of the stationary Schrödinger equation (4.59). The eigenfunctions thus read

$$\psi(x;k) = a(k) \left(\frac{1}{4}(1 - \xi^2) \right)^{-ik/2} {}_2F_1 \left(\tilde{a}, \tilde{b}; \tilde{c}; \frac{1}{2}(1 + \xi) \right) \tag{4.62}$$

where $\tilde{a} = \frac{1}{2} - ik + (V_0 + \frac{1}{4})^{1/2}$, $\tilde{b} = \frac{1}{2} - ik - (V_0 + \frac{1}{4})^{1/2}$ and $\tilde{c} = 1 - ik$ are constants depending on the model parameters and the eigenvalues. The label ${}_2F_1$ denotes the gaussian hyper-geometric function. In the limit $x \to -\infty$ $\mathrm{sech}(x) = (1 - \xi^2)^{1/2} =$

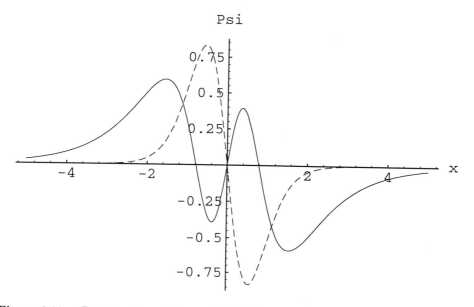

Figure 4.11.   Eigenfunctions of the modified Pöschel–Teller potential for discrete eigenvalues $n = 1$ (solid) and $n = 3$ (dashed) at $N = 4$.

$2e^x/(1 + e^{2x}) \sim 2e^x$ and the solution reduces to the form $\psi \sim a(k)e^{-ikx}$. The explicit representation in the limit $\xi \to -1$ of the solution (4.62) is given by

$$\psi(x; k) = a(k)e^{-ikx}\left(1 + \frac{\tilde{a}\tilde{b}\frac{1}{2}(1 + \xi)}{\tilde{c}} + O(\xi^2)\right).\tag{4.63}$$

The asymptotic expansion of the hyper-geometric function $_2F_1$ is done in *Mathematica* by first replacing the argument $\frac{1}{2}(1 + \xi)$ with z and then by expanding $_2F_1$ up to first order around $z = 0$.

```
Series[Hypergeometric2F1[a,b,c,z],{z,0,1}]

      a b z        2
1 + ----- + O[z]
      c
```

Hence the leading term in the asymptotic representation of the eigenfunction ψ for $x \to -\infty$ is

$$\psi \sim a(k)e^{-ikx}.\tag{4.64}$$

In the other limit $x \to +\infty$, we first transform the hyper-geometric function using the linear transformation $_2F_1(a, b, c, z) = d_2 F_1(a, b, c, 1 - z)$ yielding

$$_2F_1 \left(\tfrac{1}{2} - ik + \left(V_0 + \tfrac{1}{4} \right)^{1/2}, \tfrac{1}{2} - ik - \left(V_0 + \tfrac{1}{4} \right)^{1/2}; 1 - ik; \tfrac{1}{2}(1 + \xi) \right) \qquad (4.65)$$

$$= \left\{ \tfrac{1}{2}(1 - \xi) \right\}^{ik} {}_2F_1 \left(\tfrac{1}{2} - \left(V_0 + \tfrac{1}{4} \right)^{1/2}, \tfrac{1}{2} + \left(V_0 + \tfrac{1}{4} \right)^{1/2}; 1 - ik; \tfrac{1}{2}(1 + \xi) \right)$$

$$= \left\{ \tfrac{1}{2}(1 - \xi) \right\}^{ik} \left\{ {}_2F_1 \left(\tfrac{1}{2} - \left(V_0 + \tfrac{1}{4} \right)^{1/2}, \tfrac{1}{2} + \left(V_0 + \tfrac{1}{4} \right)^{1/2}; 1 + ik; \tfrac{1}{2}(1 - \xi) \right) \right.$$

$$* \frac{\Gamma(1 + ik)\Gamma(-ik)}{\Gamma(\tfrac{1}{2} + (V_0 + \tfrac{1}{4})^{1/2} - ik)\Gamma(\tfrac{1}{2} - (V_0 + \tfrac{1}{4})^{1/2} - ik)}$$

$$+ \left\{ \tfrac{1}{2}(1 - \xi) \right\}^{-ik} {}_2F_1 \left(\tfrac{1}{2} - \left(V_0 + \tfrac{1}{4} \right)^{1/2}, \tfrac{1}{2} + \left(V_0 + \tfrac{1}{4} \right)^{1/2}; 1 - ik; \tfrac{1}{2}(1 + \xi) \right)$$

$$\left. * \frac{\Gamma(1 - ik)\Gamma(ik)}{\Gamma(\tfrac{1}{2} + (V_0 + \tfrac{1}{4})^{1/2})\Gamma(\tfrac{1}{2} - (V_0 + \tfrac{1}{4})^{1/2})} \right\}.$$

If the potential depth is of the form $V_0 = N(N+1)$, we observe that $\tfrac{1}{2} - (V_0 + \tfrac{1}{4})^{1/2}$ is always a negative integer. Since the function Γ is singular for these points, the second term on the right hand side always vanishes. Taking this into account (4.65) reduces to

$$_2F_1 \left(\tfrac{1}{2} - ik + \left(V_0 + \tfrac{1}{4} \right)^{1/2}, \tfrac{1}{2} - ik - \left(V_0 + \tfrac{1}{4} \right)^{1/2}; 1 - ik; \tfrac{1}{2}(1 + \xi) \right) \qquad (4.66)$$

$$= \left\{ \tfrac{1}{2}(1 - \xi) \right\}^{ik} \left\{ {}_2F_1 \left(\tfrac{1}{2} - \left(V_0 + \tfrac{1}{4} \right)^{1/2}, \tfrac{1}{2} + \left(V_0 + \tfrac{1}{4} \right)^{1/2}; 1 + ik; \tfrac{1}{2}(1 - \xi) \right) \right.$$

$$\left. * \frac{\Gamma(1 + ik)\Gamma(-ik)}{\Gamma(\tfrac{1}{2} + (V_0 + \tfrac{1}{4})^{1/2} - ik)\Gamma(\tfrac{1}{2} - (V_0 + \tfrac{1}{4})^{1/2} - ik)} \right\}.$$

In the limit $x \to \infty$ the wave function ψ has the representation

$$\psi \sim e^{-ikx} + b(k)e^{ikx} \qquad (4.67)$$

where $b(k)$ is the reflection coefficient of the wave. Relation (4.67) means that an incoming wave of amplitude one is reflected by a part determined by $b(k)$.

An asymptotic expansion of the hyper-geometric function for $\xi \to 1$ consequently gives us the representation in the form

$$\psi \sim a(k) \frac{\Gamma(1 + ik)\Gamma(-ik)}{\Gamma(\tfrac{1}{2} + (V_0 + \tfrac{1}{4})^{1/2} - ik)\Gamma(\tfrac{1}{2} - (V_0 + \tfrac{1}{4})^{1/2} - ik)} e^{-ikx}. \qquad (4.68)$$

Comparing relation (4.68) with (4.67), we observe that the reflection coefficient of the wave vanishes. The transmission coefficient $a(k)$ in the case $V_0 = N(N + 1)$ takes the form

$$a(k) = \frac{\Gamma(\tfrac{1}{2} + (V_0 + \tfrac{1}{4})^{1/2} - ik)\Gamma(\tfrac{1}{2} - (V_0 + \tfrac{1}{4})^{1/2} - ik)}{\Gamma(1 + ik)\Gamma(-ik)}. \qquad (4.69)$$

A wave is free of reflection if the potential takes the form $V = V_0 \operatorname{sech}(x)$ and the depth of the potential is an integer number $V_0 = N(N+1)$.

For $V_0 = N(N+1)$, the entire calculation procedure can be activated by **AsymptoticPT[ ]** which is part of the package **AnharmonicOscillator'**. By calling **AsymptoticPT[ ]** we get the asymptotic representation of the eigenfunction in the limits $x \to \pm\infty$. The results of the expansion are contained in the global variables **w1a** and **w2a**. Function **AsymptoticPT[ ]** can also handle the case in which N is an integer. In addition to the eigenfunction, function **AsymptoticPT[ ]** delivers information about the reflection and transmission coefficients $|b|^2$ and $|a|^2$. These two characteristic properties of the scattering problem satisfy $|a|^2 + |b|^2 = 1$. **PlotPT[ ]**, which is also part of the package **AnharmonicOscillator'**, gives a graphical representation of the reflection and transmission coefficients. This function plots five curves for different k values. The range of the k values is specified as first and second arguments in the function **PlotPT[ ]**. The third argument of **PlotPT[ ]** determines the coefficient. We can choose between two types of coefficients. While "**t**" will create a plot for the transition coefficient, the "**r**" string will create the reflection plot. Two examples for $kini = 0.05$ and $kend = 0.5$ are given in figures 4.12 and 4.13. The pictures are created by **PlotPT[0.05,0.5,"r"]** and **PlotPT[0.05,0.5,"t"]** respectively.

The structure represented in figures 4.12 and 4.13 is repeated in each of the intervals $\{N, N+1 | N \geq 1\}$. Two neighboring intervals for a potential depth ranging between $V_0 = 2$ and $V_0 = 6$ ($N = 1$ and $N = 2$) are represented in figure 4.14. In this figure the reflection coefficient is shown for a range of k values by means of a surface plot. The picture is created by the sequence

```
AsymptotikPT[N,k]
Plot3D[Reflection, {N,1,3}, {k,0.05,0.75},
    AxesLabel->{"N","k","|b(k)|^2"}]
```

A collection of functions examining the anharmonic Pöschel–Teller potential is contained in the package **AnharmonicOscillator'**. Useful functions in examining the anharmonic model are **PoeschelTeller[ ]**, **AsymptoticPT[ ]** and **PlotPT[ ]**.

```
——————— File anharmon.m ———————
BeginPackage["AnharmonicOscillator'"]

Clear[AsymptoticPT,PlotPT,PoeschelTeller]

PoeschelTeller::usage= "PoeschelTeller[x_, n_, indexN_] calculates the
eigenfunction of the Poeschel Teller potential for discrete eigenvalues.
N determines the depth of the potential V0 Sech[x] by V0=N(N+1).
n fixes the state where 0 < n <= N."

w1a::usage= "The variable contains the analytic expression for the asymptotic
approximation for x -> -Infinity."
```

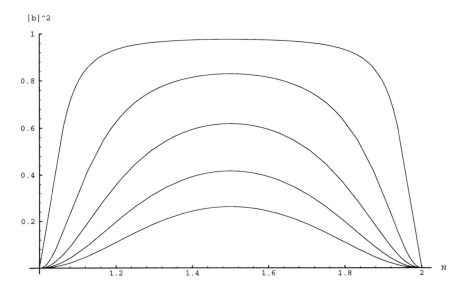

Figure 4.12. The reflection coefficient $|b|^2$ is plotted as a function of N. The ensemble of curves represent the reflection coefficient for energy values k in the interval $k \in [0.05, 0.5]$ for $N \in [1, 2]$. The top curve represents the value $k = 0.05$. The other k values > 0.05 follow below the top curve.

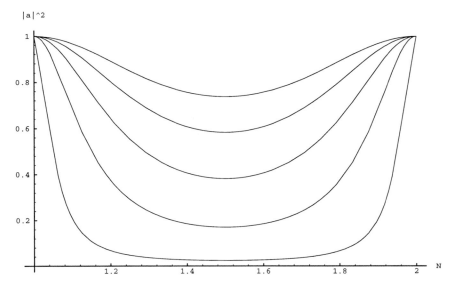

Figure 4.13. The transmission coefficient $|a|^2$ of the Pöschel–Teller potential is plotted across the depth parameter N of the potential. The energy values k are taken from the interval $k \in [0.05, 0.5]$ for $N \in [1, 2]$. The lowest curve corresponds to $k = 0.05$.

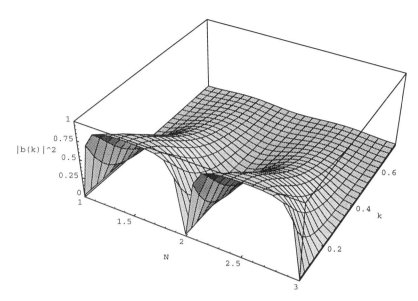

Figure 4.14. The reflection coefficient is plotted as a function of N and k. The values for the potential depth are taken from $N \in [1, 3]$ and the energy interval is $k \in [0.05, 0.75]$. We observe that the reflection coefficient decreases as the energy increases.

```
w2a::usage= "The variable contains the analytic expression for the asymptotic
approximation for x -> Infinity."

Transmission::usage= "Variable containing the expression for the transmission
coefficient. The independent variables are N and k."

Reflection::usage= "Variable containing the reflection coefficient. The
independent variables are N and k."

AsymptoticPT::usage= "AsymptoticPT[indexN_,kin_] determines the asymptotic
approximation for |x|->Infinity for the continuous case of eigenvalues in
a Poeschel Teller potential. The function yields an analytic expression for
|b(k)|^2. The variables Transmission and Reflection contain the expressions
for the transmission and the reflection coefficients. w1a and w2a contain the
approximations for x->-Infinity and x->Infinity, respectively."

PlotPT::usage= "PlotPT[kini_,kend_,type_] gives a graphical representation
of the reflection or transmission coefficient depending on the value of the
variable type. If type is set to the string r the reflection coefficient is
plotted. If type is set to t the transmission coefficient is represented. This
function creates 5 different curves."

Begin["'Private'"]

(* --- define the eigenfunctions --- *)
```

```
PoeschelTeller[x_, n_Integer, indexN_Integer] :=
  Block[{norm, integrand, xi},
     If[n <= indexN && n > 0,
(*--- eigenfunctions are the associated Legendre polynomials ---*)
     integrand = LegendreP[indexN, n, xi];
(* --- calculate the normalization constant --- *)
     norm = Integrate[integrand^2/(1-xi^2), {xi, -1, 1}];
(* --- normalize and simplify the functions --- *)
     integrand = integrand/Sqrt[norm] /. xi -> Tanh[x];
     Simplify[integrand],
(* --- error conditions --- *)
     If[indexN<n,
     Print["---  wrong argument n > N"]];
     If[n<0,
     Print["---  wrong argument n < 0"]]]
     ]


(* --- asymptotic expansion --- *)

AsymptoticPT[indexN_,kin_]:=Block[
     {rule1, rule2, wavefkt1, wavefkt2, asympt1, w1, asymt2,
      w2, akh, bkh},
(* --- replacement rules for the parameters --- *)
     rule1 = {a -> 1/2 - I k + (1/4 + V0)^(1/2),
              b -> 1/2 - I k - (1/4 + V0)^(1/2),
              c -> 1 - I k};
     rule2 = {V0 -> indexN (1 + indexN)};
     wavefkt1 = ak ((1-xi^2)/4)^(-I k/2);
     wavefkt2 = Hypergeometric2F1[a, b, c,(1+xi)/2];
(* --- asymptotic expansion for x -> -Infinity,
        equation 4.63                               --- *)
     asymt1 = Series[wavefkt2,{xi,-1,0}];
     w1= wavefkt1 Normal[asympt1] /. rule1;
     w1 = w1 /. rule2;
     w1 = w1 /. xi->Tanh[x];
     w1 = Simplify[w1];
     w1 = w1 /. Sech[x]->Exp[x];
     w1a = PowerExpand[w1];
(* --- asymptotic expansion for x -> Infinity by
        equation 4.65-4.68                          --- *)
     asymt2 = Series[wavefkt2,{xi,1,1}];
(* --- invert substitution --- *)
     w2= wavefkt1 Normal[asymt2] /. xi->Tanh[x];
(* --- eliminate higher terms --- *)
     w2 = Expand[Simplify[w2 /. -1+Tanh[x]->0]];
(* --- asymptotic behavior for Sech[] and Tanh[] --- *)
     w2 = w2 /. {Sech[x]->Exp[-x],1-Tanh[x]->Exp[-2x]};
     w2 = w2 /. rule1;
     w2 = w2 /. rule2;
     w2a = PowerExpand[w2];
(* --- determine the coefficients a[k] and b[k] --- *)
     akh = Coefficient[w2a,Exp[-I k x]] /. ak -> 1;
     bkh = Coefficient[w2a,Exp[I k x]] /. ak->1/akh;
(* --- calculate the transmission and reflection coefficient --- *)
     Transmission = 1/(akh Conjugate[akh]) /. k->kin;
     Reflection = bkh Conjugate[bkh] /. k->kin
       ]
```

```
(* --- graphical representation of the reflection and transmission
       coefficient --- *)

PlotPT[kini_,kend_,type_]:=Block[{dk,k,label},
    dk = N[(kend - kini)/5];
    k = N[kini];
    Do[
        AsymptoticPT[indexN,k];
          If[type == "r",
          p[i] = Reflection;
          label = "|b|^2",
          p[i] = Transmission;
          label = "|a|^2",];
        k  = N[k + dk],
    {i,1,5}];
    Plot[{p[1],p[2],p[3],p[4],p[5]},{indexN,1,2},AxesLabel->{"N",label},
        Prolog->Thickness[0.001]]
    ]
End[]
EndPackage[]
```

4.5 Motion in the central force field

The stationary states of a particle in a spherically symmetric potential are determined by the Schrödinger equation with the Hamiltonian operator

$$\hat{H} = -\frac{\hbar^2}{2m}\nabla^2 + V(r), \tag{4.70}$$

where $r = \sqrt{x^2 + y^2 + z^2}$ measures the distance of the particle from the origin of the potential. Using the spherical symmetry of the problem we can rewrite the Schrödinger equation in spherical coordinates

$$\left[-\frac{\hbar^2}{2m}\frac{1}{r}\frac{\partial^2}{\partial r^2}r - \frac{\hbar^2}{2mr^2}\left\{ \frac{1}{\sin\vartheta}\frac{\partial}{\partial\vartheta}\sin\vartheta\frac{\partial}{\partial\vartheta} + \frac{1}{\sin^2\vartheta}\frac{\partial^2}{\partial\phi^2} \right\} \right] \psi(r,\vartheta,\phi)$$

$$+V(r)\psi(r,\vartheta,\phi) - E\psi(r,\vartheta,\phi) = 0 \tag{4.71}$$

or in a more compact form

$$\left\{ -\frac{\hbar^2}{2m}\frac{1}{r}\frac{\partial^2}{\partial r^2}r + \frac{\hbar^2}{2mr^2}\hat{L}^2 + V(r) - E \right\}\psi = 0. \tag{4.72}$$

where $\hat{L}^2$ is the square of the angular momentum operator. Problems which can be identified by such a Hamiltonian operator are very common in physics, such as those listed below:

1. the H–atom,

2. an ion with one electron,

3. the three-dimensional harmonic oscillator,

4. the three-dimensional potential well,

5. the Yukawa particle (a shielded Coulomb potential), and

6. the free particle.

In close analogy to classical motion in a central field, we find in quantum mechanics that the angular momentum is conserved. The angular momentum is defined by

$$\boldsymbol{L} = \boldsymbol{x} \times \boldsymbol{p}. \tag{4.73}$$

Other constants of motion are the Hamiltonian, the square of the angular momentum, and the z component of the angular momentum. The related operators $\hat{H}$, $\hat{L}^2$ and $\hat{L}_z$ create a complete system of commuting operators. The solutions of the related eigenvalue problems completely determine the properties of the system. As in classical mechanics, we can take advantage of the conservation of angular momentum to reduce a three-dimensional problem to a one-dimensional one. Similarly, we can use the conservation of the angular momentum to separate the coordinates r, ϑ and ϕ in the Schrödinger equation (4.72).

The dependence of the wave function ψ on the angles ϑ and ϕ is determined by the operators $\hat{L}^2$ and $\hat{L}_z$. In spherical coordinates, we can express the z component of the angular momentum by $\hat{L}_z = -\hbar \partial_\phi$. The eigenvalues of $\hat{L}_z$, are found by solving the equation

$$-i\hbar \frac{\partial \psi(\phi)}{\partial \phi} = L_z \psi(\phi) \tag{4.74}$$

where $0 \leq \phi \leq 2\pi$. The solutions of equation (4.74) are

$$\psi(\phi) = A e^{iL_z \phi / \hbar}. \tag{4.75}$$

Since the solution (4.75) must be uniquely defined, it has to satisfy the condition

$$\psi(\phi) = \psi(\phi + 2\pi). \tag{4.76}$$

The eigenvalues $L_z/\hbar = m$ where $m = 0, \pm 1, \pm 2, \ldots$ satisfy condition (4.76). The eigenvalues of the operator $\hat{L}_z$ are thus discrete and represented by

$$L_z = \hbar m \qquad \text{where} \qquad m = 0, \pm 1, \pm 2, \ldots. \tag{4.77}$$

Since we require normalized eigenfunctions (i.e., $\int_0^{2\pi} \psi_m^* \psi_m d\phi = 1$), the normalized solutions are

$$\psi_m(\phi) = \frac{1}{\sqrt{2\pi}} e^{im\phi}. \tag{4.78}$$

A similar treatment yields the eigenvalues and eigenfunctions of the square of the angular momentum $\hat{L}^2$ from the differential equation

$$\hat{L}^2 \psi = L^2 \psi. \tag{4.79}$$

In spherical coordinates the operator $\hat{L}^2$ is represented by

$$\hat{L}^2 = -\hbar^2 \left\{ \frac{1}{\sin\vartheta} \frac{\partial}{\partial\vartheta} \left(\sin\vartheta \frac{\partial}{\partial\vartheta} \right) + \frac{1}{\sin^2\vartheta} \frac{\partial^2}{\partial\phi^2} \right\}. \tag{4.80}$$

Inserting expression (4.80) into equation (4.79), we get

$$\left\{ \frac{1}{\sin\vartheta} \frac{\partial}{\partial\vartheta} \left(\sin\vartheta \frac{\partial}{\partial\vartheta} \right) + \frac{1}{\sin^2\vartheta} \frac{\partial^2}{\partial\phi^2} + \frac{L^2}{\hbar^2} \right\} \psi(\vartheta, \phi) = 0. \tag{4.81}$$

Equation (4.81) is the defining equation of the spherical harmonics $Y_{\ell,m}$ if the eigenvalues satisfy $L^2 = \hbar^2 \ell(\ell + 1)$ with $\ell = 0, 1, 2, \ldots$

$$\left\{ \frac{1}{\sin\vartheta} \frac{\partial}{\partial\vartheta} \left(\sin\vartheta \frac{\partial}{\partial\vartheta} \right) + \frac{1}{\sin^2\vartheta} \frac{\partial^2}{\partial\phi^2} + \ell(\ell + 1) \right\} Y_{\ell,m}(\vartheta, \phi) = 0. \tag{4.82}$$

The eigenvalues of $\hat{L}^2$ are determined by the quantum numbers $\ell = 0, 1, 2, \ldots$. Their related eigenfunctions are the spherical harmonics $Y_{\ell,m}$ of order ℓ. Comparing the structure of the eigenfunctions of the harmonic oscillator to that of the eigenfunctions of the angular momentum $\hat{L}^2$, we observe that in the case of $\hat{L}^2$ with eigenvalues $L^2 = \hbar^2 \ell(\ell+1)$ there are $2\ell+1$ eigenfunctions $Y_{\ell,m}$. The eigenfunctions $Y_{\ell,m}$, however, are different in the second quantum number m, which is known as the magnetic quantum number. For a fixed value of L^2, m counts the different projections on the z axis. If we determine ℓ, we find different values for m

$$m = 0, \pm 1, \pm 2, \ldots, \pm \ell,$$

and limited to the range $-\ell \leq m \leq \ell$. For the proof of the above relations, we refer the reader to the book by Cohen-Tannoudji [4.6].

The complete representation of the spherical harmonics for positive m is

$$Y_{\ell,m}(\vartheta, \phi) = \frac{1}{\sqrt{2\pi}} e^{im\phi} (-1)^m \sqrt{\frac{(2\ell + 1)(\ell - m)!}{2(\ell + m)!}} \sin^m \vartheta \, P_\ell^m(\cos\vartheta). \tag{4.83}$$

$P_\ell^m(x)$ denotes the m-th associated Legendre function of order ℓ. In case of negative quantum numbers m, we use the relation

$$Y_{\ell,-m}(\vartheta, \phi) = (-1)^m Y_{\ell,m}^*(\vartheta, \phi). \tag{4.84}$$

If we use the representation of the spherical harmonics given by relation (4.83), it is easy to show that the $Y_{\ell,m}$ are also eigenfunctions of the operator $\hat{L}_z$. By a simple calculation we find

$$- i\hbar \frac{\partial}{\partial \phi} Y_{\ell,m}(\vartheta, \phi) = \hbar m Y_{\ell,m}(\vartheta, \phi). \tag{4.85}$$

We now can state that the spherical harmonics are eigenfunctions of both the z component of the angular momentum operator and the square of the angular momentum operator. The corresponding eigenvalues are

$$L^2 = \hbar^2 \ell(\ell + 1) \quad \text{and} \quad L_z = \hbar m. \tag{4.86}$$

The spherical harmonics are accessed in *Mathematica* by the function **Spherical-HarmonicY**[ ]. The Legendre polynomials are available using **LegendreP**[ ].

So far we have determined the eigenfunctions depending on ϑ and ϕ. Separating the angular terms from the radial part of the wave function, we get the representation

$$\psi(r, \vartheta, \phi) = h(r) Y_{\ell,m}(\vartheta, \phi). \tag{4.87}$$

Relation (4.87) used with equation (4.72) allows the derivation of a determining equation for the radial part $h(r)$ of the wave function ψ. The wave function separates because the coordinate system of our problem is separable. The radial function $h(r)$ is dependent on the energy E, the quantum number ℓ and on the potential energy $V(r)$. Consequently, the radial part of the wave function is independent of m: in a spherical potential there are no distinguishing directions breaking the symmetry.

Inserting relation (4.87) into the Schrödinger equation (4.72) and using our above results for the angular momentum, we get after substituting $u(r) = rh(r)$ the eigenvalue problem for the radial part of the wave function

$$-\frac{\hbar^2}{2m} \frac{d^2 u(r)}{dr^2} + \left\{ V(r) + \frac{\hbar^2 \ell(\ell+1)}{2mr^2} \right\} u(r) = Eu(r). \tag{4.88}$$

$u(r) = rh(r)$ is substituted since for $r \to 0$, the function $h(r)$ has to be finite (i.e., $u(r) \to 0$ for $r \to 0$). Note that in equation (4.88) all parameters are known except for potential $V(r)$. For the following discussion we assume that the potential $V(r)$ represents a Coulomb interaction of the two particles,

$$V(r) = -\frac{Ze^2}{r}. \tag{4.89}$$

This type of potential applies to the hydrogen and hydrogen–like atoms where $Z = 1$ as well as to ionized atoms like He^+, Li^{++}, etc.

The stationary states of an electron in a Coulomb potential result from the eigenvalue equation

$$\frac{d^2 u(r)}{dr^2} + \left\{ \frac{2mE}{\hbar^2} + \frac{2mZe^2}{\hbar^2 r} - \frac{\ell(\ell+1)}{r^2} \right\} u(r) = 0. \tag{4.90}$$

To carry out our calculation, it is convenient to introduce scaled variables

$$\varrho = \frac{r}{a} \quad \text{and} \quad \epsilon = \frac{E}{E_0} \tag{4.91}$$

where $a = \hbar^2/(me^2) \approx 5.29 \cdot 10^{-11}$m is the Bohr radius and $E_0 = e^2/(2a) = me^4/\hbar^2 \approx 13.5$eV the ionization energy of the hydrogen atom. The Schrödinger equation (4.90) is thus represented by

$$\left\{ \frac{d^2}{d\varrho^2} + \epsilon + \frac{2Z}{\varrho} - \frac{\ell(\ell+1)}{\varrho^2} \right\} u(\varrho) = 0. \tag{4.92}$$

We restrict our calculations to the case of bound states characterized by negative energy values. To find appropriate representations of a solution for $u(r)$, we examine the limits $r \to 0$ and $r \to \infty$. The function $u(r)$ is either given by a polynomial in ϱ $u_\ell(\varrho) = r^\alpha(1 + a_1\varrho + a_2\varrho^2 + \ldots)$ or by an exponential relation $u_\ell(\varrho) = Ae^{-\gamma\varrho} + Be^{\gamma\varrho}$ where $\gamma^2 = -\epsilon$. The results of these expressions are conditions for the parameters α and B which satisfy $\alpha = \ell + 1$, $B = 0$. Using these results both expressions are reducible to

$$u_\ell(\varrho) = \varrho^{\ell+1} e^{-\gamma\varrho} f(\varrho). \tag{4.93}$$

Substituting expression (4.93) into equation (4.92) and using $x = 2\gamma\varrho$, we get the standard form of Kummer's differential equation

$$xf'' + (2(\ell+1) - x)f' - \left(\ell + 1 - \frac{Z}{\gamma}\right) f = 0, \tag{4.94}$$

where primes denote differentiation with respect to x. The solutions of equation (4.94) are confluent hypergeometric functions $({}_1F_1)$

$$f_\ell(\varrho) = c\,{}_1F_1\left(\ell + 1 - \frac{Z}{\gamma}, 2\ell + 2; 2\gamma\varrho\right). \tag{4.95}$$

To satisfy the normalization condition, series (4.95) must terminate at a finite order. This restriction induces the quantization of the energy values by

$$\ell + 1 - \frac{Z}{\gamma} = -n_r \qquad \text{with} \qquad n_r = 0, 1, 2, \ldots . \tag{4.96}$$

The solution of equation (4.96) with respect to γ delivers

$$\gamma = \frac{Z}{n_r + \ell + 1} \tag{4.97}$$

or, by replacing $\gamma^2 = -\epsilon$, yields energy values $\epsilon = -Z^2/(n_r + \ell + 1)^2$ to be

$$E = -\frac{E_0 Z^2}{(n_r + \ell + 1)^2} = -\frac{E_0 Z^2}{n^2}. \tag{4.98}$$

The quantum number n is the principal quantum number determined by the radial quantum number n_r ($n_r = 0, 1, 2, \ldots$) and the angular quantum number ℓ ($\ell = 0, 1, 2, \ldots$). The wave function of the electron in the Coulomb potential is given by

$$\psi_{n,\ell,m}(\varrho,\vartheta,\phi) = N_{n,\ell}\varrho^\ell e^{Z\varrho/n} {}_1F_1\left(\ell+1-n, 2\ell+2; 2\frac{Z}{n}\varrho\right) Y_{\ell,m}(\vartheta,\phi) \qquad (4.99)$$

where $N_{n,\ell}$ is the normalization constant

$$N_{n,\ell} = \frac{1}{(2\ell+1)!}\sqrt{\frac{(n+\ell)!}{2n(n-\ell-1)!}}\left(\frac{2Z}{n}\right)^{\ell+3/2}. \qquad (4.100)$$

The radial part of the wave function $h(\varrho)$ consists of

$$h_{n,\ell}(\varrho) = N_{n,\ell}\varrho^\ell e^{-Z\varrho/n} {}_1F_1\left(\ell+1-n, 2\ell+2; 2\frac{Z}{n}\varrho\right). \qquad (4.101)$$

Since the first argument in the hypergeometric function is a negative integer, the function ${}_1F_1$ in the radial part reduces to a polynomial known as a Laguerre polynomial. In *Mathematica* the Laguerre polynomials are denoted by **LaguerreL[ ]**. One useful parameter of the radial wave function is $n_r = n - \ell - 1$. This parameter counts the nodes of the eigenfunction along the horizontal axis. This behavior is shown in figure 4.15 for $n = 3$ and $\ell = 0, 1, 2$. Figure 4.15 is created by

```
Plot[{Radial[r,3,0,1],Radial[r,3,1,1],Radial[r,3,2,1]},
     {r,0,25},AxesLabel->{"r","h"},Prolog->Thickness[0.001]]
```

The function **Radial[ ]** used in the **Plot[ ]** function is part of the package **CentralField'**. This package also contains **Angle[ ]** for the angular part of the wave function. The definition of **Angle[ ]** is in some ways redundant since *Mathematica* accounts for the angular part of the wave function under the name **SphericalHarmonicY[ ]**. However, we separately define the angular part of the wave function to show how relations (4.83) and (4.84) are expressed in terms of *Mathematica*.

The above wave function is applied to representations of orbitals of an atom or a molecule. Chemists, for example, work with molecular orbital theory to describe the binding of atoms. This theory makes extensive use of the angular wave functions $Y_{\ell,m}$. In order to describe the binding of a molecule it is necessary to use a linear combination of the angular parts of the wave function. We create such a superposition of the $Y_{\ell,m}$'s by the function **Orbital[ ]** which is part of the package **CentralField'** . **Orbital[ ]** creates sums and differences of the spherical harmonics in the form

$$w(\vartheta,\phi) = |Y_{\ell,m} \pm Y_{\ell,-m}|^2. \qquad (4.102)$$

Relation (4.102) represents the probability of finding an electron within a certain domain of the angular part of the space. In figures 4.16–4.19, we have plotted some particular examples for orbitals.

Figures 4.16–4.19 show an inner view of the orbitals for a certain range of ϕ. Similar pictures for other quantum numbers are created by the superposition of the angular wave functions $Y_{\ell,m}$ with the help of **Angle[ ]** . The figures of the orbitals are created by the function sequence **AnglePlot[Orbital[l,m,theta,-phi,"plus"],theta,phi]** .

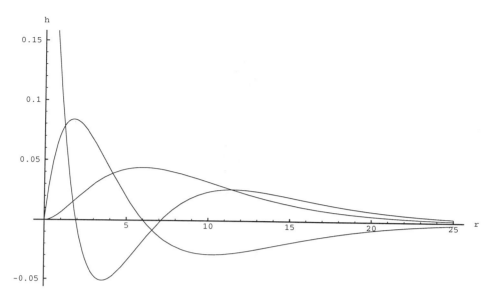

Figure 4.15.   Radial part h of the wave function for $n = 3$ and $\ell = 0, 1, 2$.

─────────────────────── File centralf.m ───────────────────────

```
BeginPackage["CentralField'"]
Clear[Radial,Angle,AnglePlot,Orbital];

Radial::usage = "Radial[ro_, n_, l_, Z_] calculates the radial representation
of the eigenfunctions for an electron in the Coulomb potential. The numbers
n and l are the quantum numbers for the energy and the angular momentum
operator. Z specifies the number of charges in the nucleus. The radial
distance between the center and the electron is given by ro."

Angle::usage = "Angle[theta_, phi_, l_, m_] calculates the angular part of
the wave function for an electron in the Coulomb potential. The numbers L
and m denote the quantum numbers for the angular momentum operator. Theta
and phi are the angles in the spherical coordinate system."

Orbital::usage= "Orbital[theta_,phi_,l_,m_,type_String]   calculates the
superposition of two wave functions for the quantum numbers m_l = +m and
m_l = -m. The variable type allows the creation of the sum or the difference
of the wave functions. The string values of type are either plus or minus."

AnglePlot::usage = "AnglePlot[pl_,theta_,phi_] gives a graphical
representation of the function contained in pl. The range of representation
is Pi <= phi < 5 Pi/2 and 0 < theta < Pi. Theta is measured with
respect to the vertical axis. This function is useful for ploting the orbitals
and the angular part of the eigenfunction."

(* --- define global variables --- *)
```

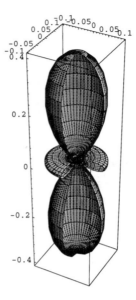

Figure 4.16.   Angular part of the wave function for $\ell = 2$ and $m = 0$.

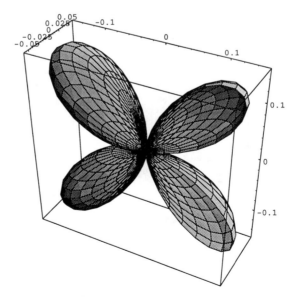

Figure 4.17.   Orbital for the quantum numbers $\ell = 2$ and $m = \pm 1$ formed from the difference $|Y_{2,1} - Y_{2,-1}|^2$.

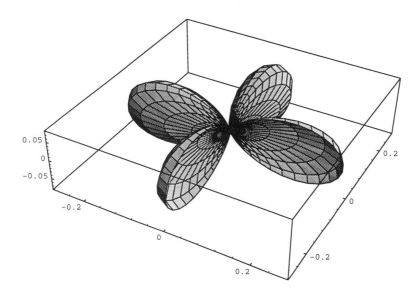

Figure 4.18.   A plot of the sum of the wave functions with quantum numbers $\ell = 2$ and $m = \pm 2$.

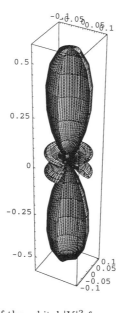

Figure 4.19.   Representation of the orbital $|Y|^2$ for quantum numbers $\ell = 3$ and $m = 0$.

```
theta::usage
phi::usage
ro::usage
n::usage
l::usage
m::usage

Begin["'Private'"]

(* --- radial part of the eigenfunctions in the Coulomb potential --- *)

Radial[ro_, n_, l_, Z_] :=
  Block[{norm, hnl},
(* --- normalization --- *)
    norm = (Sqrt[(n + 1)!/(2 n (n - l - 1)!)] ((2 Z)/n)^(l + 3/2))/
           (2 l + 1)!;

(* --- definition of the wave function --- *)
    hnl = norm ro^l Exp[-((Z ro)/n)]
      Hypergeometric1F1[l + 1 - n, 2 l + 2, (2 Z ro)/n]]


(* --- angular part of the eigenfunctions in the Coulomb field --- *)

Angle[theta_, phi_, l_, m_] :=
    Block[{norm, legendre, x, angle , m1, result},
      m1 = Abs[m];
(* --- normalization --- *)
      norm = (-1)^m1 Sqrt[((2 l + 1) (l - m1)!)/(2 (l + m1)!) ]/
            Sqrt[2 Pi];
(* --- eigenfunctions --- *)
      legendre = Sin[theta]^m1 D[LegendreP[l, x], {x, m1}];
      legendre = legendre /. x -> Cos[theta];
(* --- consider the cases m > 0 and m < 0 --- *)
      If[m >= 0, angle  = Exp[I m phi],
                 angle  = (-1)^m1 Exp[-(I m1 phi)]];
(* --- normalized eigenfunction --- *)
      result = norm legendre angle
      ]


(* -- create orbitals --- *)

Orbital[theta_,phi_,l_,m_,type_String]:=
    Block[{norm, m1, rule, wave, wave2},
      m1 = Abs[m];
(* --- replacement rule for the exponential function --- *)
      rule = {E^(Complex[0, a_] (x_.)) ->
                Cos[x Abs[a]] + I Sign[a] Sin[x Abs[a]]};
(* --- distinguish different cases --- *)
      If[m1 >= 1,
            If[type == "plus",
(* --- sum of the wave functions for a fixed m --- *)
            wave = Expand[Angle[theta,phi,l,m1] +
                          Angle[theta,phi,l,-m1] /. rule],
(* --- difference of the wave function for a fixed m --- *)
            wave = Expand[Angle[theta,phi,l,m1] -
                          Angle[theta,phi,l,-m1] /. rule]
```

```
            ];
       wave2 = wave^2;
 (* --- normalization of the superposition --- *)
       norm = Integrate[wave2, {phi,0,2Pi}, {theta,0,Pi}];
       wave2 = Expand[wave2/Abs[norm]],
          wave = Angle[theta,phi,1,m1]^2]
       ]

(* --- graphical representation of the angular part --- *)

AnglePlot[pl_,theta_,phi_]:= Block[{},
(* --- theta is measured with respect to the vertical axis --- *)
    ParametricPlot3D[{-pl Sin[theta] Cos[phi],-pl Sin[theta] Sin[phi],
                      pl Cos[theta]},
                     {phi,Pi,5Pi/2},{theta,0,Pi},
                     PlotRange->All,
                     PlotPoints->{40,40}]
                 ]
End[]
EndPackage[]
```

4.6 Exercises

1. Examine the spectrum of the eigenvalues for a potential well with different depths. Give a graphical representation of the eigenvalues depending on different depths.

2. Determine the wave functions for different eigenvalues for the potential well by using the methods discussed in Section 4.1.

3. Check the relation $|a|^2 + |b|^2 = 1$ for the anharmonic oscillator.

4. Re-examine the Pöschel–Teller problem and study the expectation values $\langle x^n \rangle$ given by

$$\langle x^n \rangle = \int \psi^* x^n \psi dx$$

for different values of n.

5. Plot the radial part of the wave function of the hydrogen atom for different quantum numbers n and ℓ. Examine the influence of the charge Z.

6. Create a graphical representation of the f orbital for the Europium atom.

5

Nonlinear dynamics

5.1 The Korteweg-de Vries equation

Weak nonlinear waves can be described by an integro–differential equation of the form

$$u_t - uu_x + \int_{-\infty}^{\infty} K(x - \xi)u_\xi(\xi, t)d\xi = 0. \tag{5.1}$$

The dispersive behavior of the waves is contained in kernel K. We obtain the dispersion relation K by a Fourier transform of the related phase velocity $c(k) = \omega(k)/k$ by

$$K(x) = \frac{1}{2\pi} \int_{-\infty}^{\infty} c(k)e^{-kx}dx \tag{5.2}$$

where $\omega(k)$ is the dispersion relation of the wave. The Korteweg-de Vries (KdV) equation was first derived at the end of the 19th century to describe water waves in shallow channels. Experimental data of the dispersion relation in such channels show that the square of the phase velocity is expressible by the relation

$$c^2(k) = \frac{g}{k}\tanh(kh), \tag{5.3}$$

where h is the mean depth of the channel measured from the undisturbed surface of the water and g is the acceleration of gravity of the earth. For waves with large wavelengths, we observe that the argument of tanh is small. Thus we can use a Taylor expansion to approximate the phase velocity by

$$
\begin{aligned}
c(k) &= \sqrt{\frac{g}{k}\tanh(kh)} \\
&\approx \sqrt{gh}\left\{1 - \frac{h^2}{6}k^2 + O(k^4)\right\}.
\end{aligned} \tag{5.4}
$$

As a consequence, the kernel K given in equation (5.2) is represented by an expansion in the form

$$
\begin{aligned}
K(x) &= \frac{1}{2\pi} \int_{-\infty}^{\infty} \sqrt{gh} \left\{ 1 - \frac{h^2}{6} k^2 \right\} e^{ikx} dk \\
&= \sqrt{gh} \left\{ \delta(x) + \frac{h^2}{6} \delta''(x) \right\}.
\end{aligned}
\tag{5.5}
$$

where $\delta(x)$ is the Dirac delta function and the primes denote derivatives with respect to the argument. If we consider these relations in our original equation of motion (5.1), we get

$$
u_t - uu_x + \sqrt{gh} \int_{-\infty}^{\infty} \left\{ \delta(x - \xi) + \frac{h^2}{6} \delta''(x - \xi) \right\} u_\xi(\xi, t) d\xi = 0
$$

$$
u_t - uu_x + \sqrt{gh} \left\{ u_x + \frac{h^2}{6} u_{xxx} \right\} = 0.
\tag{5.6}
$$

Transforming equation (5.6) to a moving coordinate system by $X = x + vt$ for $v = -\sqrt{gh}$, scaling the time t and the wave amplitude u by $\tau = h^2 t/6$ and $\tilde{u} = u/h$ respectively results in a standard representation of the KdV equation

$$
u_t - 6uu_x + u_{xxx} = 0.
\tag{5.7}
$$

In equation (5.7), we use the original variables to denote the transformed quantities.

In the following we derive the KdV equation following the reasoning above and using *Mathematica*. To derive the KdV it is necessary to expand the phase velocity in Fourier space. An inverse transformation from Fourier space to the real coordinate space gives us the representation of the phase velocity in real coordinates. The back transformation from the Fourier space is obtained by the *Mathematica* function **InverseFourierTransform**[ ] contained in the standard package **Calculus'FourierTransform'**. Note that Dirac's δ–function is involved in the calculation of equation (5.5). We need to be careful when using the δ–function in any calculation with *Mathematica* since the rules governing its use have not been completely defined. We first need to define rules to handle the integration of derivatives of the δ–function. These rules are collected in the list **DeltaRule1** in package **KdVEquation'**. Note that each rule is accompanied by a condition marked by /;. If the individual condition is satisfied, the rule is applied; if not, the rule is ignored. The rules defined in **DeltaRule1** are applied to the integration of the function **Equation**[ ] as long as no further simplification occurs. The function **ReplaceRepeated**[ ] (//.) performs this multiple application of rules.

```
──────────────────── File kdvequat.m ────────────────────

BeginPackage["KdVEquation'"]

Needs["Calculus'FourierTransform'"]

Clear[DeltaRule1, DeltaRule2, c, disperse, K, Equation]
```

```
Equation::usage ="Equation[n_] calculates the evolution equation up to order
n."

(* -- declare global variables --- *)

x::usage
t::usage
u::usage
h::usage
g::usage

Begin["'Private'"]
(* --- rules to handle Dirac's delta function --- *)

DeltaRule1 = {
(* --- basic rules for the delta function --- *)

Integrate[c_. + expr_. Delta[a_. t_ + b_.], t_] :>
  Integrate[c, t] + If[True, -expr/a /. t-> -b/a] /;
    FreeQ[{a, b}, t] && FreeQ[expr,Delta],
Integrate[c_. + expr_. Delta[a_. t_ + b_.], {t_, t1_, t2_}] :>
  Integrate[c, {t, t1, t2}] +
    Simplify[If[True, -expr/a /. t-> -b/a]] /;
    FreeQ[{a, b}, t] && FreeQ[expr,Delta],

(* --- rules for higher derivatives of the delta function --- *)

Integrate[c_. +
  expr_. Derivative[n_][Delta][a_. t_ + b_.], t_] :>
  Integrate[c, t] + If[True, (-1)^(n+1)
    D[expr,{t,n}]/a /. t-> -b/a]
    /;
    FreeQ[{a, b}, t] && FreeQ[expr,Delta],
Integrate[c_. +
  expr_. Derivative[n_][Delta][a_. t_ + b_.], {t_, t1_, t2_}] :>
  Integrate[c, {t, t1, t2}] +
    Simplify[If[True, (-1)^(n+1) D[expr,{t,n}]/a /.
    t-> -b/a]] /;
    FreeQ[{a, b}, t] && FreeQ[expr,Delta],

Integrate[(expr1_. + expr2_.) expr3_., {t_, t1_, t2_}] :>
  Integrate[expr1 expr3, {t, t1, t2}] +
  Integrate[expr2 expr3, {t, t1, t2}]
  /; FreeQ[expr3,Delta]
};

(* --- propagation velocity (dispersion relation) --- *)

c[k_] := Block[{g,h},
      Sqrt[g Tanh[k h]/k] ]

(* --- expansion of the propagation velocity --- *)

disperse[k_,n_]:= Block[{},
      Normal[Series[c[k],{k,0,n}]]
      ]
```

```
(* --- inverse Fourier transformation of the dispersion relation --- *)

K[xi_,n_] := Block[{k, itrafo, dis, t},
        dis = disperse[k,n];
        itrafo = Simplify[2 Pi
        InverseFourierTransform[dis,k,t]];
        itrafo = itrafo /. t -> x-xi]

(* --- derive the equation of motion --- *)

Equation[n_]:= Block[{gl},
        gl = Integrate[K[xi,n] D[u[xi,t],xi],
                        {xi,-Infinity,Infinity}] //.
                        DeltaRule1;
        gl = Simplify[gl];
        gl = D[u[x,t],t] - u[x,t] D[u[x,t],x] + gl
        ]
End[]
EndPackage[]
```

Using the function **Equation[ ]** of the package **KdVEquation'**, we derive the KdV equation applying the theoretical procedure as discussed above. If we specify the approximation order in **Equation[ ]**, we get the following result

```
Equation[3]

   (0,1)                    (1,0)
 u     [x, t] - u[x, t] u      [x, t] +

                 (1,0)          2 (3,0)
     Sqrt[g h] (6 u     [x, t] + h  u     [x, t])
 >   ---------------------------------------------
                      6
```

5.2 Solution of the Korteweg-de Vries equation

In this section we derive the analytical solutions of the KdV equations using certain initial and boundary conditions. The KdV equation is given by

$$u_t - 6uu_x + u_{xxx} = 0 \qquad \text{with} \qquad t > 0 \quad \text{and} \quad -\infty < x < \infty \qquad (5.8)$$

and the initial condition is $u(x, t = 0) = u_0(x)$. We assume natural boundary conditions; i.e., the solution of the KdV equation (5.8) is assumed to vanish sufficiently fast at $|x| \to \infty$. To arrive at our solution, we use the inverse scattering

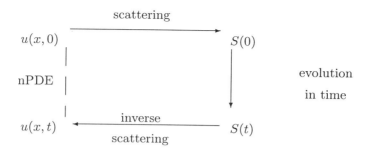

Figure 5.1.   Solution procedure of the inverse scattering transform.

theory (IST). This procedure is closely related to its linear counterpart, the Fourier transform (FT). In section 4.1 we used the Fourier transform technique to construct solutions of the Schrödinger equation. In addition to its methodical connection with IST and FT, both IST and FT are also logically related to the Sturm–Liouville problem. The main difference between IST and FT is that the Fourier transform is only capable of solving linear problems while the IST can also be applied to nonlinear differential equations. The solution steps for the inverse scattering transform are summarized as follows (see figure 5.1).

i) The starting point is a set of nonlinear partial differential equations (nPDE) for a certain initial condition $u(x,0)$.

ii) By a scattering process, we get the scattering data $S(0)$ at the initial time $t = 0$ from the initial data.

iii) Since the characteristic data of the scattering process is related to a linear problem, we can determine the time evolution of the scattering data for the asymptotic behavior $|x| \to \infty$.

iv) The inverse scattering process gives us the solution $u(x,t)$. The inverse scattering process is closely related to a linear integro–differential equation, the Marchenko equation, well known in the theory of scattering.

Using these four steps in the solution process, we get a large number of solutions. The most prominent solutions contained in this set are for solitons and multi–solitons. We note that the solution process discussed so far is not only applicable to the KdV equation but also delivers solutions for more complicated equations. A collection of equations solvable by IST is given by Calogero & Degasparis [5.1]. Note that the IST procedure is not applicable to all nonlinear initial value problems. There exists, however, a set of equations for which the IST procedure works very well. One of these equations is the KdV equation which is a completely integrable equation.

As mentioned above, the starting point of the IST is the initial condition $u(x, 0) = u_0(x)$. In close analogy to the example discussed in the chapter on quantum mechanics (section 4.4), we examine here a scattering problem with the scattering potential $u(x, 0) = u_0(x)$. To calculate the scattering data $S(0)$ we consider the related Sturm–Liouville problem in the form

$$\psi_{xx} + (\lambda - u_0(x))\psi(x) = 0 \qquad -\infty < x < \infty, \tag{5.9}$$

where λ represents the eigenvalue. The time–independent scattering data is derived from the asymptotic behavior of the wave function ψ. Our treatment of equation (5.9) is closely analogous to our calculations in quantum mechanics. The asymptotic behavior of the wave function is given by

$$\psi(x; k) \sim \begin{cases} e^{-ikx} + b(k)e^{ikx} & \text{for} \qquad x \to \infty \\ a(k)e^{-ikx} & \text{for} \qquad x \to -\infty \end{cases} \tag{5.10}$$

where $\lambda > 0$ and $k = \lambda^{1/2}$ refer to the case of a continuous spectrum, and where

$$\psi_n(x) \sim c_n e^{-\kappa_n x} \qquad \text{for} \qquad x \to \infty \quad n = 1, 2, \ldots, N \tag{5.11}$$

for $\lambda < 0$ and $\kappa_n = (-\lambda)^{1/2}$ refer to the case of discrete eigenvalues. The characteristic data of the scattering process is the set of reflection and transmission indices $b(k)$ and $a(k)$ and the normalization constant c_n. This set of data is called the scattering data $S(0)$ and is collected in a list $S(0) = \{a(k), b(k), c_n\}$. The listed data support the theory. The measurable quantities in a scattering process are the reflection and transmission coefficients $b(k)$ and $a(k)$. The question from the experimental point of view is how the measurable quantities can be used to derive the interaction potential. Theoretically, the answer is given by Marchenko [5.2]. He demonstrated that knowledge of the scattering data and eigenvalues of the Sturm–Liouville problem are sufficient to reconstruct the potential of the scattering process by a linear integral equation of the form

$$K(x, z) + M(x + z) + \int_x^\infty K(x, y)M(y + z)dy = 0, \tag{5.12}$$

where M is defined by the scattering data by

$$M(x) = \sum_{n=1}^N c_n^2 e^{-\kappa_n x} + \frac{1}{2\pi} \int_{-\infty}^\infty b(k)e^{ikx} dk. \tag{5.13}$$

The solution $K(x, z)$ of the integral equation (5.12) delivers the representation of the potential $u_0(x)$

$$-2\frac{d}{dx}K(x, x) = u_0(x). \tag{5.14}$$

Knowing the scattering data, we are able to reconstruct the potential $u_0(x)$ by means of the Marchenko equation (5.12).

Another aspect of solving the KdV equation is how time influences the scattering. Up to now we have only considered the stationary characteristics of the scattering process. We now consider not only the initial condition $u = u(x, t = 0)$ in the scattering process but also the full time–dependent behavior of the solution $u(x, t)$. We assume that the time–dependent potential $u(x, t)$ in the Sturm–Liouville problem satisfies the natural boundary conditions requiring that for $|x| \to \infty$ the solution vanishes sufficiently fast enough. In all expressions the time variable t is considered as a parameter. Because of the parametric dependency of the Sturm–Liouville problem on t, we expect that all spectral data also depend on t. We assume the eigenvalues $\lambda = \lambda(t)$ to include a time–dependence in the Sturm–Liouville problem which, in this case, reads

$$\psi_{xx}(\lambda(t) - u(x;t))\psi(x;t) = 0 \tag{5.15}$$

where $u(x, t)$ satisfies the KdV equation (5.8). Differentiation of equation (5.15) with respect to x as well as with respect to t gives us

$$\psi_{xxx} - u_x\psi + (\lambda - u)\psi_x = 0 \tag{5.16}$$
$$\psi_{xxt} + (\lambda_t - u_t)\psi + (\lambda - u)\psi_t = 0. \tag{5.17}$$

By introducing the expression

$$R(x, t) = \psi_t + u_x\psi - 2(u + 2\lambda)\psi_x, \tag{5.18}$$

we find that the current $\psi_x R - \psi R_x$ satisfies the relation

$$\frac{\partial}{\partial x}(\psi_x R - \psi R_x) = \lambda_t \psi^2 \tag{5.19}$$

which connects the time derivative of the eigenvalues λ to the gradient of the current. To derive this relation we have used equations (5.16) and (5.17) as well as the KdV equation (5.8) itself.

If the eigenvalues λ of the Sturm–Liouville problem are discrete ($\kappa_n = (-\lambda)^{1/2}$), an integration of equation (5.19) with respect to x yields

$$0 = \psi_x R - \psi R_x|_{-\infty}^{\infty} = \lambda_t \int_{-\infty}^{\infty} \psi^2 dx. \tag{5.20}$$

Since the wave function ψ and its derivatives vanish for $|x| \to \infty$, the left hand side of equation (5.20) is gone. Normalizing ψ by $\int_{-\infty}^{\infty} \psi^2 dx = 1$ results in

$$\frac{d\kappa_n^2}{dt} = 0 \qquad \text{or} \qquad \kappa_n = const. \tag{5.21}$$

We therefore have an isospectral problem. We now can use equation (5.18) to determine directly the normalization constants c_n. On the other hand, u and ψ vanish for $x \to \infty$. Using the asymptotic representation of the eigenfunctions ψ, we find, with the help of

$$\psi_n(x;t) \sim c_n(t)e^{-\kappa_n x} \tag{5.22}$$

and the asymptotic form (5.18)

$$\frac{dc_n}{dt} - 4\kappa_n^3 c_n = 0. \tag{5.23}$$

Integrating this expression gives

$$c_n(t) = c_n(0)e^{4\kappa_n^3 t} \qquad n = 1, 2, \ldots, N \tag{5.24}$$

where $c_n(0)$ are the normalization constants of the time–independent Sturm–Liouville problem. Following these steps, we see how the discrete part of the spectral data follows from the time–independent eigenvalue problem.

The continuous part of the spectral data is derived by an analogous procedure. The integration of relation (5.19) with respect to x produces the continuous part of the eigenvalues

$$\psi_x R - \psi R_x = g(t;k). \tag{5.25}$$

The asymptotic representation of the eigenfunctions is now

$$
\begin{aligned}
\psi(x;t,k) &\sim a(k;t)e^{-ikx} && \text{for} && x \to -\infty \\
\psi(x;t,k) &\sim e^{-ikx} + b(k;t)e^{ikx} && \text{for} && x \to \infty
\end{aligned}
\tag{5.26}
$$

In the limiting case of $x \to -\infty$ we find by using (5.18)

$$R(x;t,k) \sim \left(\frac{da}{dt} + 4ik^3 a\right)e^{-ikx} \tag{5.27}$$

and thus we obtain

$$\psi_x R - \psi R_x \to 0 \qquad \text{for} \qquad x \to -\infty. \tag{5.28}$$

This relation allows a further integration which results in

$$R = h(t;k)\psi. \tag{5.29}$$

Using (5.29) we get the expression

$$\frac{da}{dt} + 4ik^3 a = ha. \tag{5.30}$$

The corresponding relations for $x \to \infty$ are expressed by

$$\frac{db}{dt}e^{ikx} + 4ik^3\left(e^{-ikx} - be^{ikx}\right) = h\left(e^{-ikx} + be^{ikx}\right). \tag{5.31}$$

Since the trigonometric functions are linearly independent functions we can write

$$\frac{db}{dt} - 4ik^3 b = hb \tag{5.32}$$

$$h = 4ik^3. \tag{5.33}$$

Equation (5.30) is thus reducible to

$$\frac{da}{dt} = 0. \tag{5.34}$$

A simultaneous integration of equations (5.34) and (5.32) gives

$$a(k;t) = a(k;0) \tag{5.35}$$

$$b(k;t) = b(k;0)e^{8ik^3 t}. \tag{5.36}$$

For times $t > 0$, we obtain a time–dependent reflection index $b(k;t)$ and a constant transmission rate $a(k;t)$.

The complete set of scattering data (discrete plus continuous data) for the time–dependent scattering problem of the KdV equation is summarized as follows

$$S(t) = \left\{ c_n(t) = c_n(0)e^{4\kappa_n^3 t}, a(k;0), b(k;t) = b(k;0)e^{i8k^3 t} \right\}. \tag{5.37}$$

The assumption of a time–dependent potential is reflected in the scattering data through both the time–dependent normalization constants c_n in the discrete spectrum and the time–dependent reflection coefficients b in the continuous spectrum.

To complete the solution process of the inverse scattering transform, we need to take into account the time–dependence of the scattering data in Marchenko's integral equation. Since time appears only as a parameter in the relations of the scattering data, we can use the expression from the stationary part of the scattering process and extend it to obtain the equations of the time–dependent scattering. The time–dependent potential and the solution of the KdV equation follow from the time–dependent Marchenko equation. The spectral characteristics are contained in the M–term. If we generalize relation (5.13) for the time–dependent case of spectral data, we get

$$M(x;t) = \sum_{n=1}^{N} c_n(0)^2 e^{8\kappa_n^3 t - \kappa_n x} + \frac{1}{2\pi} \int_{-\infty}^{\infty} b(k;t)e^{i(8k^3 t - kx)} dk. \tag{5.38}$$

The original Marchenko equation then transforms to

$$K(x,z;t) + M(x+z;t) + \int_{x}^{\infty} K(x,y;t)M(y+z;t)dy = 0. \tag{5.39}$$

The solution of the KdV equation follows from

$$u(x,t) = -2\frac{\partial}{\partial x}K(x,x;t).\tag{5.40}$$

In principle, equation (5.40) gives the solution for the KdV equation provided the spectral data are known. However, deriving the spectral data is not simple, even for the KdV equation. Calculating the general solution of the Marchenko equation is a second problem in the solution process. This situation is similar to the Fourier technique for which the inverse transformation is at times unrecoverable. Given a spectral density $A(k)$, it is sometimes impossible to analytically invert the representation from Fourier space into real space. However, since our main problem is the application of the IST, we show in the following section that the IST can be successfully applied to the solution of the KdV equation.

5.2.1 Soliton solutions of the Korteweg-de Vries equation

In the previous section, we saw how nonlinear initial value problems can be solved using the inverse scattering method. In this section we construct the solution for a specific problem. As an initial condition we choose the potential in the Sturm–Liouville problem to be $u_0(x) = -V_0 \operatorname{sech}^2(x)$. We already discussed this type of potential in section 4.4 when examining the Pöschel-Teller problem. We observed there that the reflection index $b(k)$ vanishes if the amplitude of the potential is given by $V_0 = N(N+1)$ with N an integer. In our discussion of solutions for the KdV equation, we restrict our considerations to this case.

We assume that $N = 1$. The initial condition is thus reduced to $u_0(x) = -2\operatorname{sech}^2(x)$. The related Sturm–Liouville problem (5.9) reads for this specific case

$$\psi_{xx} + (\lambda + 2\operatorname{sech}^2(x))\psi = 0.\tag{5.41}$$

Equation (5.41) is identical to equation (4.57) with $V_0 = 2$. We know from the quantum mechanical treatment of the problem that in this case the corresponding eigenfunctions are given by the associated Legendre polynomials $P_1^1(x) = \operatorname{sech}(x)/\sqrt{2}$. The corresponding eigenvalue is $\kappa_1 = 1$. The normalization constant follows from the normalization condition $\int_{-\infty}^{\infty} \psi^2 dx = 1$. According to our considerations in the previous section we can immediately write down the time history of the normalization constant c_1 as

$$c_1(t) = \sqrt{2}e^{4t}.\tag{5.42}$$

Since we are dealing with a reflectionless potential $(b(k) = 0)$ we can write the M term of the Marchenko equation as

$$M(x;t) = 2e^{8t-x}.\tag{5.43}$$

The Marchenko equation itself reads

$$K(x, z; t) + 2e^{8t-(x+z)} + 2\int_x^\infty K(x, y; t)e^{8t-(y+z)}dy = 0. \qquad (5.44)$$

Solutions of equation (5.44) are derivable by a separation expression for the function K in the form $K(x, z; t) = \mathcal{K}(x; t)e^{-z}$. Substituting this expression into equation (5.44) gives us the relation

$$\mathcal{K}(x; t) + 2e^{8t-x} + 2\mathcal{K}(x; t)\int_x^\infty e^{-2y}dy = 0. \qquad (5.45)$$

We have thus reduced an integral equation to an algebraic relation for $\mathcal{K}$. The solution of (5.45) is given by

$$\mathcal{K}(x; t) = -\frac{2e^{8t-x}}{1 + e^{8t-2x}}. \qquad (5.46)$$

The unknown $K(x, z; t)$ is thus represented by

$$K(x, z; t) = -\frac{2e^{8t-x}}{1 + e^{8t-2x}}e^{-z}. \qquad (5.47)$$

In fact, the solution of the KdV can be obtained using equation (5.39) to derive the time–dependent potential $u(x, t)$ from K

$$
\begin{aligned}
u(x, t) &= 2\frac{\partial}{\partial x}\left(\frac{2e^{8t-2x}}{1 + e^{8t-2x}}\right) \\
&= -2\operatorname{sech}^2(x - 4t). \qquad (5.48)
\end{aligned}
$$

This type of solution is known as the soliton solution of the KdV. It was first derived at the end of the 19th century by Korteweg and de Vries. The solution itself describes a wave with constant shape and constant propagation velocity $v = 4$ moving to the right. By choosing the amplitude, we derive one solution out of an infinite set of solutions for the KdV equation. In the following, we discuss more complicated cases where two and more eigenvalues have to be taken into account for the calculation.

To demonstrate how IST can be applied to more complicated situations, consider the case with an initial condition $u_0(x) = -6\operatorname{sech}^2(x)$. The difference between this case and the case discussed above appears to be minor. However, as we shall see, the difference in the solutions is significant. The selected initial condition corresponds to a Pöschel-Teller potential with a depth of $N = 2$. Recalling the discussion of section 4.4, the eigenvalues are given by $\kappa_1 = 1$ and $\kappa_2 = 2$. The corresponding eigenfunctions are

$$\psi_2^1 = \left(\frac{3}{2}\right)^{1/2}\tanh(x)\operatorname{sech}(x) \qquad (5.49)$$

$$\psi_2^2 = \frac{\sqrt{3}}{2}\operatorname{sech}^2(x). \qquad (5.50)$$

The normalization constants c_1 and c_2 for this case are given by

$$c_1 = \sqrt{6} \qquad \text{and} \qquad c_2 = 2\sqrt{3}. \tag{5.51}$$

The time history of c is determined by

$$
\begin{aligned}
c_1(t) &= \sqrt{6}e^{4t} & (5.52) \\
c_2(t) &= 2\sqrt{3}e^{32t}. & (5.53)
\end{aligned}
$$

In close analogy to $N = 1$, we get the M terms of the Marchenko equation by using relation (5.38) in the form

$$M(x;t) = 6e^{8t-x} + 12e^{64t-2x}. \tag{5.54}$$

The Marchenko equation itself is given by

$$
\begin{aligned}
K(x, z; t) &+ 6e^{8t-(x+z)} + 12e^{64t-2(x+z)} \\
&+ \int_x^\infty K(x, y; t) \left\{ 6e^{8t-(y+z)} + 12e^{64t-2(y+z)} \right\} dy = 0.
\end{aligned} \tag{5.55}
$$

We obtain the solution of equation (5.55) in the form

$$K(x, z; t) = \mathcal{K}_1(x; t)e^{-z} + \mathcal{K}_2(x; t)e^{-2z} \tag{5.56}$$

by again using a separation expression for K. In the general case of N eigenvalues, we can use the expression

$$K(x, z; t) = \sum_{n=1}^{N} \mathcal{K}_n(x; t)e^{-nz} \tag{5.57}$$

to reduce the integral equation to an algebraic relation. Since e^{-z} and e^{-2z} are linearly independent functions, we can derive from equation (5.55) the following system of equations

$$\mathcal{K}_1 + 6e^{8t-x} + 6e^{8t}\left(\mathcal{K}_1 \int_x^\infty e^{-2y}dy + \mathcal{K}_2 \int_x^\infty e^{-3y}dy \right) = 0 \tag{5.58}$$

$$\mathcal{K}_2 + 12e^{64t-2x} + 12e^{64t}\left(\mathcal{K}_1 \int_x^\infty e^{-3y}dy + \mathcal{K}_2 \int_x^\infty e^{-4y}dy \right) = 0. \tag{5.59}$$

Integrating equations (5.58) and (5.59), we get a linear system of equations for the unknowns $\mathcal{K}_i$

$$
\begin{pmatrix} 1 + 3e^{8t-2x} & 2e^{8t-3x} \\ 4e^{64t-3x} & 1 + 3e^{64t-4x} \end{pmatrix} \begin{pmatrix} \mathcal{K}_1 \\ \mathcal{K}_2 \end{pmatrix} = \begin{pmatrix} -6e^{8t-x} \\ -12e^{64t-2x} \end{pmatrix}. \tag{5.60}
$$

For cases with $N > 2$, we get a general system of equations which is

$$A\mathcal{K} = B \tag{5.61}$$

where

$$A_{n,m} = \delta_{mn} + \frac{c_m^2(0)}{m+n} e^{8m^3 t - (m+n)x} \tag{5.62}$$

and

$$B_n = -c_n^2(0) e^{8n^3 t - nx}. \tag{5.63}$$

The final solution reads

$$u(x,t) = -2 \frac{\partial^2}{\partial x^2} \log |A|. \tag{5.64}$$

Equation (5.64) is the general representation of the solution for the KdV equation. For the specific case with $N = 2$, we get

$$
\begin{aligned}
\mathcal{K}_1(x,t) &= 6(e^{72t-5x} - e^{8t-x})/D(x,t) \tag{5.65} \\
\mathcal{K}_2(x,t) &= -12(e^{64t-2x} + e^{72t-4x})/D(x,t). \tag{5.66}
\end{aligned}
$$

The determinant $D(x,t) = \det A = |A|$ of (5.60) is

$$D(x,t) = 1 + 3e^{8t-2x} + 3e^{64t-4x} + e^{72t-6x}. \tag{5.67}$$

The solution of the KdV equation then reads

$$
\begin{aligned}
u(x,t) &= -2 \frac{\partial}{\partial x} \left(\mathcal{K}_1 e^{-x} + \mathcal{K}_2 e^{-2x} \right) \\
&= -12 \frac{3 + 4 \cosh(2x - 8t) + \cosh(4x - 64t)}{\{3 \cosh(x - 28t) + \cosh(3x - 36t)\}^2}. \tag{5.68}
\end{aligned}
$$

This type of solution is called a bi–soliton solution in the theory of inverse scattering. To make the term soliton more understandable, we examine the behavior of the solution (5.68) in a certain time domain. Since the KdV equation is invariant with respect to a Galilean transformation, we can use $t < 0$ in our calculations. A sequence of time steps illustrating (5.68) is presented in figures 5.2–5.4. In order to give the impression of a wave packet, we have plotted the negative amplitude of the solution u in these figures. Initially, there are two separated peaks. As time passes, the two humps overlap and form a single peak at time $t = 0$, which represents the initial solution $u_0(x) = -6 \operatorname{sech}^2(x)$. For times $t > 0$, we observe that the single peak located at $x = 0$ splits into two peaks with differing amplitudes. We observe that wave packets with larger amplitudes split from those with smaller amplitudes. Larger wave packets travel faster than smaller ones. If we compare the soliton movement before and after the collision of pulses, we observe during the scattering process that neither the shapes nor the velocities of the pulses changes. The term soliton originates from its insensitivity to any variance in the scattering process. This phenomenon was first observed by Zabusky and Kruskal [5.5]. Another characteristic of solitons is that larger pulses travel faster while smaller pulses move

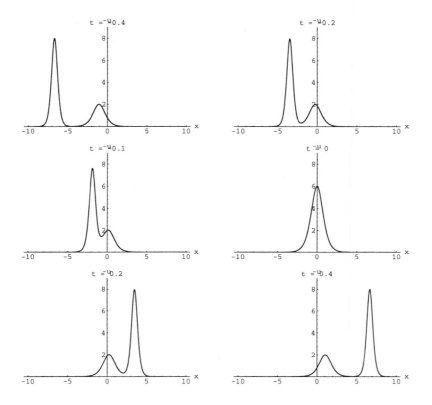

Figure 5.2.   Soliton solution of the KdV equation. The initial condition is
$$u(x,0) = -6\,\text{sech}^2(x).$$

more slowly. This means that larger pulses will overtake smaller ones during the
time history of motion. We can understand this time history by examining the
propagation velocity with respect to the amplitude of the solitons.

From figure 5.2, we note that for times $|t| \to \infty$ the shape of the solitons
remains stable. As already mentioned, the shape of the pulses is recovered in a
scattering process. However, the phase of the pulses does not stay continuous. It
smoothly changes at the interaction of the solitons. A two soliton scattering is
pictured in figure 5.3, created with **ContourPlot[ ]**. We observe in this plot that
smaller packets retard while larger ones advance.

The *Mathematica* functions needed to create the figures for the soliton movement are collected in the package **KDVAnalytic'**. The function needed to plot
the solitons is **Soliton[ ]**, and a graphical representation of an N–soliton solution
is obtained by using the function **PlotKdV[ ]**. An example of a quartic soliton
solution is given in figure 5.4, created by calling **PlotKdV[-0.5,0.5,0.02,4]**. The
four pictures created in the time domain ranging from $t = -0.5$ up to $t = 0.5$ in

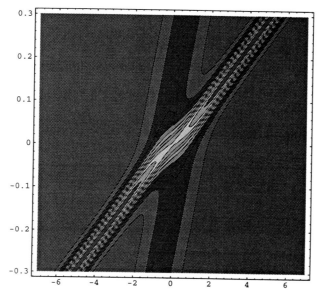

Figure 5.3. Contour plot of the bi–soliton solution. The space coordinate x is plotted horizontally while time t is plotted vertically. We can clearly detect the discontinuity of the phase in the contour plot at $t = 0$. The gap occurs in the spatial direction.

steps of $\Delta t = 0.02$ are collected in one picture by using **Show[ ]** in connection with **GraphicsArray[ ]**.

```
————————————————— File kdvanaly.m —————————————————

BeginPackage["KdVAnalytic`"]

Clear[PlotKdV,c2,Soliton];

Soliton::usage = "Soliton[x_,t_,N_] creates the N soliton solution of the KdV
equation."

PlotKdV::usage = "PlotKdV[tmin_,tmax_,dt_,N_] calculates a sequence of
pictures for the N soliton solution of the KdV equation. The time interval
of the representation is [tmin,tmax]. The variable dt measures the length
of the time step."


(* --- global variables --- *)

x::usage
t::usage


Begin["`Private`"]

(* --- squares of the normalization constants c_n --- *)
```

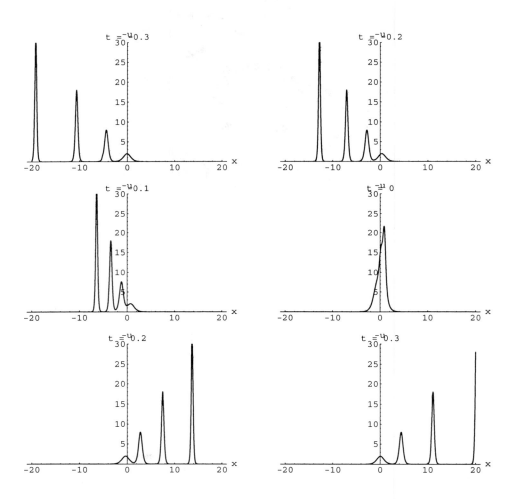

Figure 5.4.   Time series for a quartic–soliton solution. The given time points are
$t = -0.5, 0.00001, 0.3$.

```
c2[n_, N_] := Block[{h1,x},
    h1 = LegendreP[N, n, x]^2/(1-x^2);
    h1 = Integrate[h1, {x, -1, 1}]]

(* --- N soliton solution --- *)

Soliton[x_,t_,N_] :=
  Block[{cn,A,x,t,deltanm,u},
(* --- calculate normalization constants --- *)
    cn = Table[c2[i, N], {i, 1, N}];
(* --- create the coefficient matrix A --- *)
    A = Table[
              If[n==m, deltanm = 1, deltanm=0];
              deltanm + (cn[[m]] Exp[8 m^3 t - (m + n) x])/(m + n),
              {m, 1, N}, {n, 1, N}];
(* --- determine the solution --- *)
    u = -2 D[Log[Det[A]],{x,2}];
    u = Expand[u];
    u = Factor[u]]

(* --- time series of the N soliton solution --- *)

PlotKdV[tmin_,tmax_,dt_,N_]:=Block[{p1,color,u},
(* --- create the N soliton --- *)
    u = Soliton[x,t,N];
(* --- plot the N soliton --- *)
    Do[
        If[t>0,color=RGBColor[0,0,1],color=RGBColor[1,0,0]];
        Plot[-u,{x,-20,20},PlotRange->{0,15},
            AxesLabel->{"x","-u(x,t)"},
            DefaultColor->Automatic,
            PlotStyle->{{Thickness[1/170],color}}],
        {t,tmin,tmax,dt}]]
End[]
EndPackage[]
```

5.3 Conservation laws of the Korteweg-de Vries equation

Conservation laws such as the conservation of energy are central quantities in physics. The conservation of angular momentum is equally important to quantum mechanics as it is to classical mechanics. Conservation laws imply the existence of invariant quantities, when applied to scattering of molecules. The Boltzmann equation is an example, as the particle density remains constant, since particles are neither created nor destroyed.

Denoting the macroscopic particle density with $\varrho(x,t)$ and the streaming velocity with $v(x,t)$, we can express the conservation law in the differential form of a continuity equation

$$\partial_t \varrho(x,t) + \partial_x(\varrho v) = 0. \tag{5.69}$$

Assuming that the current $j = \varrho v$ vanishes for $|x| \to \infty$ and integrating over the domain $x \in (-\infty, \infty)$, we get for the density ϱ the relation

$$\frac{d}{dt} \left(\int_{-\infty}^{\infty} \varrho dx \right) = -\varrho v|_{-\infty}^{\infty} = 0, \tag{5.70}$$

and thus

$$\int_{-\infty}^{\infty} \varrho dx = const. \tag{5.71}$$

Equation (5.71) expresses the conservation of mass although the density ϱ follows the time history in accordance with equation (5.69). The simple idea of mass conservation in fluids can also be transformed to more general situations. If we write down for a general density T and its corresponding current J a continuity equation such as

$$\partial_t T + \partial_x J = 0, \tag{5.72}$$

we find the related conservation law. To extend the formulation of the general continuity equation to nonlinear partial differential equations, we assume that T and J depend on $t, x, u, u_x, u_{xx}, \ldots$ but not on u_t. If we retain the assumption that $J(x \to \pm\infty) \to 0$, then equation (5.72) can be integrated over all space as was done for equation (5.69), getting

$$\frac{d}{dt} \int_{-\infty}^{\infty} T dx = 0 \tag{5.73}$$

or

$$\int_{-\infty}^{\infty} T dx = const. \tag{5.74}$$

The quantity defined by (5.74) is an integral of motion in the theory of nonlinear PDEs.

As an example, we consider the KdV equation

$$u_t - 6uu_x + u_{xxx} = 0. \tag{5.75}$$

The KdV equation already takes the form of a continuity equation. $T_1 = u$ is the density and $J = u_{xx} - 3u^2$ is the current. If the density T is integrable and $\partial_x J$ vanishes at the points $x = \pm\infty$, we can write

$$\int_{-\infty}^{\infty} u(x, t) dx = const. \tag{5.76}$$

Equation (5.76) must be satisfied for all solutions of the KdV equation satisfying the conditions listed above. However, not all solutions of the KdV equation satisfy the asymptotic relations. For example, the conservation laws do not apply to periodic solutions of the KdV equation.

Another conserved quantity can be obtained if equation (5.75) is multiplied by u. In this case

$$\partial_t \left(\frac{1}{2}u^2 \right) + \partial_x \left\{ uu_{xx} - \frac{1}{2}u_x^2 - 2u^3 \right\} = 0. \tag{5.77}$$

The second conserved quantity is given by $T_2 = u^2$ which directly integrates into

$$\int_{-\infty}^{\infty} u^2 dx = const. \tag{5.78}$$

This notation holds for solutions vanishing sufficiently rapidly at $|x| \to \infty$. The physical interpretation of these equations is that relation (5.76) represents conservation of mass and that equation (5.78) represents conservation of momentum (compare also section 5.1). We have thus derived two conserved quantities by simple manipulations of the KdV equation. The question now is whether we can derive other conserved quantities from the KdV and how these quantities are related to each other. This question was first discussed by Miura, Gardner & Kruskal [5.3]. They observed that there are a large number of conserved quantities for the KdV equation. They discovered that in fact there exists an infinite number of conserved quantities for the KdV equation. For example

$$T_3 = u^3 + \frac{1}{2}u_x^2 \tag{5.79}$$

$$T_4 = 5u^4 + 10uu_x^2 + u_{xx}^2. \tag{5.80}$$

T_3 can be identified as the energy density. The higher densities T_n for $n > 3$ have no physical interpretation in terms of energy, momentum, etc. Other conserved quantities are obtained algorithmically. In the following, we show how Miura *et al* constructed the infinite hierarchy of constants of motion.

5.3.1 Derivation of conservation laws

Miura *et al* [5.3] made an important step in understanding the phenomenon of invariants in nonlinear PDEs. The tool they invented is a transformation vehicle which linearizes the nonlinear PDE. Today, this tool is known as the Miura transformation of the KdV equation to the modified KdV equation (mKdV),

$$v_t - 6v^2 v_x + v_{xxx} = 0. \tag{5.81}$$

By transforming the field v to the field u according to

$$u(x,t) = v^2(x,t) + v_x(x,t), \tag{5.82}$$

solutions of (5.81) are also solutions of the KdV equation. The Miura transformation $v = \psi(x,t)/\psi_x(x,t)$ connects the KdV equation with its related Sturm–Liouville problem. The Miura transformation (5.82) is primarily used for the construction of conservation laws. If, for example, we replace field v in (5.82) by

$$v = \frac{1}{2\epsilon} + \epsilon w, \tag{5.83}$$

where ϵ is an arbitrary parameter, we get the Miura transformation for w in the form

$$u = \frac{1}{4}\frac{1}{\epsilon^2} + w + \epsilon^2 w^2 + \epsilon w_x. \tag{5.84}$$

If we additionally assume the Galilean invariance for u to be $(\tilde{u} = u + \lambda)$, we can simplify the relation (5.84) to the form

$$u = w + \epsilon w_x + \epsilon^2 w^2. \tag{5.85}$$

This transformation connecting w with u is called a Gardner transformation. Substituting the transformation (5.85) into the KdV equation (5.75) gives us

$$
\begin{aligned}
u_t - 6uu_x + u_{xxx} &= w_t + \epsilon w_{xt} + 2\epsilon^2 ww_t \\
&\quad -6(w + \epsilon w_x + \epsilon^2 w^2)(w_x + \epsilon w_{xx} + 2\epsilon^2 ww_x) \\
&\quad +w_{xxx} + \epsilon w_{xxxx} + 2\epsilon^2 (ww_x)_{xx} \\
&= \left(1 + \epsilon\frac{\partial}{\partial x} + 2\epsilon^2 w\right)\left\{w_t + 6(w + \epsilon^2 w^2)w_x + w_{xxx}\right\}.
\end{aligned}
\tag{5.86}
$$

As is the case for the Miura transformation, u is a solution of the KdV equation and thus w is also a solution of the KdV equation

$$w_t - 6(w + \epsilon^2 w^2)w_x + w_{xxx} = 0. \tag{5.87}$$

If we set the parameter to be $\epsilon = 0$, equation (5.87) reduces to the KdV equation. For this case, the Gardner transformation yields the identity transformation $u = w$. The Gardner transformation is closely related to a continuity equation of the form

$$\partial_t w + \partial_x (w_{xx} - 3w^2 - 2\epsilon^2 w^3) = 0. \tag{5.88}$$

Thus we get

$$\int_{-\infty}^{\infty} w\,dx = const., \tag{5.89}$$

i.e., another conserved quantity. To construct the conservation laws of the KdV equation by an algorithm, we use the parameter ϵ. The important aspect of this operation is that for $\epsilon \to 0$, w converges to u. For this reason we expand field w as a power series in ϵ

$$w(x,t;\epsilon) = \sum_{n=0}^{\infty} \epsilon^n w_n(x,t). \tag{5.90}$$

From equation (5.89) it follows

$$\int_{-\infty}^{\infty} w\,dx = \sum_{n=0}^{\infty} \epsilon^n \int_{-\infty}^{\infty} w_n\,dx = const. \tag{5.91}$$

or

$$\int_{-\infty}^{\infty} w_n\,dx = const. \qquad for \qquad n = 0,1,2,\ldots. \tag{5.92}$$

The expansion of the Gardner transformation (5.85) yields

$$\sum_{n=0}^{\infty} \epsilon^n w_n = u - \epsilon \sum_{n=0}^{\infty} \epsilon^n w_{nx} - \epsilon^2 \left(\sum_{n=0}^{\infty} \epsilon^n w_n \right)^2. \tag{5.93}$$

The conserved quantities resulting from the first terms of this expansion are

$$w_0 = u \tag{5.94}$$
$$w_1 = -w_{0x} = -u_x \tag{5.95}$$
$$w_2 = -w_{1x} - w_0^2 = u_{xx} - u^2 \tag{5.96}$$
$$w_3 = -w_{2x} - 2w_0 w_1 = -(u_{xx} - u^2)_x + 2uu_x. \tag{5.97}$$

The quantities w_1 and w_3 are given by total differentials and thus provide new information on the conservation laws.

Since the construction of the invariants of motion follows from a completely algorithmic procedure, *Mathematica* can be used to derive the higher densities of the conservation laws. Indeed, a calculation by hand immediately shows us that a manual approach is very cumbersome. However, *Mathematica* can do all the calculations for us.

The algorithm to derive the conserved densities starts out from a power series expansion of the field w. The comparison of equal powers of ϵ in equation (5.93) gives us the expressions for the w_n's. If we replace the w_n's by the w_{n-1}'s, we get a representation of function u. The steps used to carry out the calculation are summarized in the package **KdVIntegrals'**. The **Gardner[ ]** function activates our calculation of conserved quantities. Given an integer as an argument, **Gardner[ ]** creates the first n conserved densities. These densities are collected in a list. Applying **Integrate[ ]** to the result of **Gardner[ ]**, all even densities result in an integral of motion. Results of a calculation with $n = 6$ are

```
Gardner[6]

             (1,0)              2     (2,0)
{u[x, t], -u      [x, t], -u[x, t]  + u      [x, t],

              (1,0)            (3,0)
>    4 u[x, t] u      [x, t] - u      [x, t],

             3     (1,0)     2                (2,0)           (4,0)
>    2 u[x, t]  - 5 u      [x, t]  - 6 u[x, t] u      [x, t] + u      [x, t],

              2  (1,0)            (1,0)      (2,0)
>    -16 u[x, t]  u      [x, t] + 18 u      [x, t] u      [x, t] +

              (3,0)            (5,0)
>    8 u[x, t] u      [x, t] - u      [x, t],

              4               (1,0)      2          2  (2,0)
>    -5 u[x, t]  + 50 u[x, t] u      [x, t]  + 30 u[x, t]  u      [x, t] -

           (2,0)  2    (1,0)      (3,0)
>    19 u      [x, t]  - 28 u      [x, t] u      [x, t] -

              (4,0)            (6,0)
>    10 u[x, t] u      [x, t] + u      [x, t]}
```

After integrating the list we obtain

```
                                                       2        (1,0)
{Integrate[u[x, t], x], -u[x, t], -Integrate[u[x, t] , x] + u      [x, t],

              2  (2,0)                         3    (1,0)     2
>    2 u[x, t]  - u      [x, t], Integrate[2 u[x, t]  + u      [x, t] , x] -

              (1,0)            (3,0)
>    6 u[x, t] u      [x, t] + u      [x, t],

                  3
     -16 u[x, t]          (1,0)     2           (2,0)           (4,0)
>    ------------ + 5 u      [x, t]  + 8 u[x, t] u      [x, t] - u      [x, t],
          3

                     4               (1,0)     2  (2,0)     2
>    Integrate[-5 u[x, t]  - 10 u[x, t] u      [x, t]  - u      [x, t] , x] +

              2  (1,0)            (1,0)      (2,0)
>    30 u[x, t]  u      [x, t] - 18 u      [x, t] u      [x, t] -

              (3,0)            (5,0)
>    10 u[x, t] u      [x, t] + u      [x, t]}
```

```
─────────────── File kdvinteg.m ───────────────
BeginPackage["KdVIntegrals'"]

Clear[Gardner];

Gardner::usage = "Gardner[N_] calculates the densities of the integrals of
motion for the KdV equation using Gardner's method. The integrals are
determined up to the order N."

(* --- global variables --- *)

u::usage
x::usage
t::usage

Begin["'Private'"]

Gardner[N_] :=
            Block[{expansion,eps,x,t,sublist ={},list1={},list2},
            list2=Table[1, {i,1,N+1}];

(* --- representation of a Gardner expansion  --- *)
            expansion = Expand[
                    Sum[eps^n w[x,t,n] - eps^(n+1) D[w[x,t,n],x],
                    {n,0,N}] -
                    eps^2 (Sum[eps^n w[x,t,n], {n,0,N}])^2 - u[x,t]
                    ];
(* --- compare coefficients --- *)
            Do[AppendTo[list1,
                        Expand[Coefficient[expansion,eps,i]-w[x,t,i]]],
            {i,0,N}];
            list2[[1]] = -list1[[1]];
(* --- define replacements and application of the replacements --- *)
            Do[sublist={};
                Do[AppendTo[sublist,w[x,t,i]->list2[[i+1]] ],
                    {i,0,N}];
                AppendTo[sublist,D[w[x,t,n],x]->D[list2[[n+1]],x]];
                list2[[n+2]] = list1[[n+2]] /. sublist,
                {n,0,N-1}];
                list2
                ]
End[]
EndPackage[]
```

5.4 Numerical solution of the Korteweg-de Vries equation

Our considerations of the solutions of the KdV equations have so far been restricted to reflectionless potentials, and thus we have used a special type of potential (Pöschel–Teller potential) in the analytic calculations. In this section we examine solutions of the KdV equation for arbitrary potentials $u(x, 0)$. For an arbitrary potential $u(x, 0)$, we cannot expect the reflection coefficient to be $b(k) = 0$. For a reflectionless potential, we solve the Marchenko equation by a separation expression. For $b(k) \neq 0$, however, there is no analytic procedure available to solve the Marchenko equation. In this case the KdV equation can be solved numerically. There are several procedures for finding numerical solutions of the KdV equation. An overview of the various integrating methods is given by Taha & Ablowitz [5.4].

Nonlinear evolution equations are solvable by a pseudo spectral method or by difference methods. With respect to the difference methods, there are several versions of the standard Euler method known as leap–frog and Crank–Nicolson procedures. For our numerical solution of the KdV equation, we use the leap–frog procedure as developed by Zabusky & Kruskal [5.5]; see also Crandall [5.6].

All of the difference methods represent the continuous solution $u(x, t)$ for discrete points in space and time. In the process of discretization, the space and time coordinates are replaced by $x = mh$ and $t = nk$. $m = 0, 1, \ldots, M$, $n = 0, 1, 2, \ldots$ h and k determine the step lengths in the spatial and temporal directions. Since the x–domain of integration is restricted to an interval of finite length, we choose $h = 2\pi/M$ for the step length in the x–direction. The continuous solution $u(x, t)$ is approximated for each integration step by $u(x, t) = u_m^n$; i.e., steps h and k have to be chosen properly to find convergent solutions (see the following equations).

All discretization procedures differ in the representation of their derivatives. The main challenge of the discretization procedure is to find the proper representation of the needed derivatives. Errors are inevitable in this step and we have to settle for an approximate solution. Various representations of the derivatives give us a varying degree of accuracy for the representation of the solution. The leap–frog method of Zabusky & Kruskal approximates derivatives in the KdV equation

$$u_t - 6uu_x + u_{xxx} = 0 \tag{5.98}$$

by the formula

$$u_m^{n+1} = u_m^{n-1} + \frac{6k}{3h}(u_{m+1}^n + u_m^n + u_{m-1}^n)(u_{m+1}^n - u_{m-1}^n)$$

$$- \frac{k}{h^3}(u_{m+2}^n - 2u_{m+1}^n + 2u_{m-1}^n - u_{m-2}^n). \tag{5.99}$$

The first term on the right hand side of (5.99) represents the first derivative with respect to time. The second term gives a representation of the nonlinearity in the KdV equation. The last term in the sum of the right hand side describes the dispersion term of third order in the KdV. The main advantage of the Zabusky & Kruskal procedure is the conservation of mass in the integration process $\sum_{m=0}^{M-1} u_m^n$. Another aspect of this discretization procedure is the representation of nonlinearity by $\frac{1}{3}(u_{m+1}^n + u_m^n + u_{m-1}^n)$. In this representation the energy is conserved up to second order

$$\frac{1}{2}\sum_{m=0}^{M-1}(u_m^n)^2 - \frac{1}{2}\sum_{m=0}^{M-1}(u_m^{n-1})^2 = O(k^3) \qquad \text{for} \qquad k \to 0, \tag{5.100}$$

if u is periodic or vanishes sufficiently rapidly at the integration end points. Since the Zabusky & Kruskal procedure is a second order method in the time domain, we face the problem of specifying the initial conditions for the terms u_m^n and u_m^{n-1}. This problem can be solved if we use as a first step of integration an Euler procedure given by

$$
\begin{aligned}
u_m^{n+1} &= u_m^n + \frac{6k}{3h}(u_{m+1}^n + u_m^n + u_{m-1}^n)(u_{m+1}^n - u_{m-1}^n) \\
&\quad - \frac{k}{h^3}(u_{m+2}^n - 2u_{m+1}^n + 2u_{m-1}^n - u_{m-2}^n).
\end{aligned}
\tag{5.101}
$$

To find stable solutions for this integration process, we have to choose the time and space steps appropriately. If we assume linear stability of the solution procedure, we have to take the following relation into account

$$k \le \frac{h^3}{4 + h^2|u|} \tag{5.102}$$

where $|u|$ denotes the maximum magnitude of u. The process of integration includes the following steps:

1. create the initial conditions

2. execute the first step of the integration by applying the simple Euler procedure using relations (5.101)

3. iterate the following steps by using equation (5.99)

4. create a graphical representation of the results for equal time intervals.

The above four steps for integrating the KdV equation are contained in the package **KdVNumeric'**. **KdVNIntegrate[]** activates the integration process. **KdVNIntegrate[]** needs steps h and k, the number of points used in the x–domain, and the initial solution for $t = 0$ as input parameters. Results of an integration with the initial condition $u(x,0) = -6\operatorname{sech}(x)$ are given in figure 5.5. As we know from our analytical considerations in the previous section, we expect a bi–soliton solution. Choosing a larger amplitude in the initial condition $u(x,0) = -10\operatorname{sech}(x)$, we get two solution components. In addition to the soliton properties, we observe a radiation solution in figure 5.6. The radiation part of the solution moves in the opposite direction to that of the soliton and decreases in time.

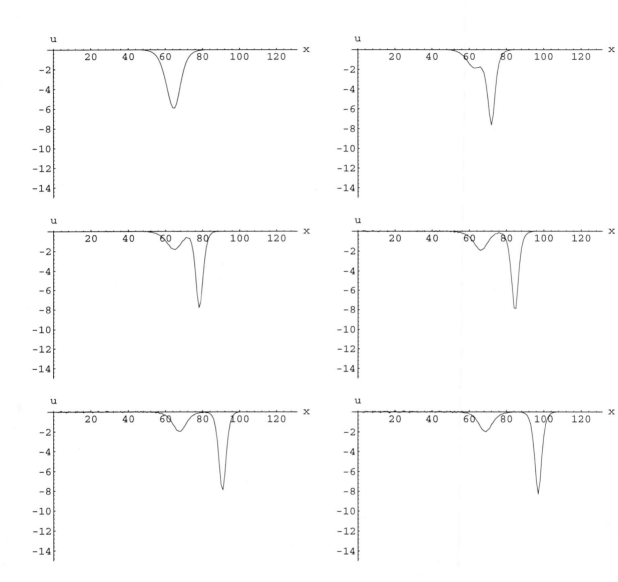

Figure 5.5.   Numerical solution of the KdV equation for the initial condition
$u(x, 0) = -6\,\mathrm{sech}(x)$. The time points shown from left to right and top to bottom are
$t = \{0, 0.16, 0.32, 0.64\}$. The calculation is based on 128 points in the x–domain corresponding to
a step size of $h = 0.2$. The steps in the time domain are $k = 0.002$.

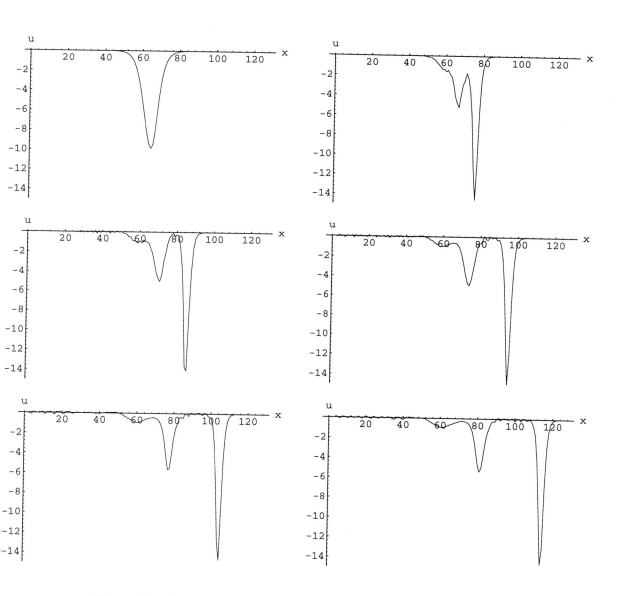

Figure 5.6. Numerical solution of the KdV equation for the initial condition
$u(x,0) = -10\,\mathrm{sech}(x)$. The time points shown from left to right and top to bottom are
$t = \{0, 0.16, 0.32\}$. The calculation is based on 128 points in the x–domain with a step size of
$h = 0.2$.

```
────────────────────── File kdvnumer.m ──────────────────────
BeginPackage["KdVNumeric`"]

Clear[KdVNIntegrate]

KdVNIntegrate::usage = "KdVNIntegrate[initial_,dx_,dt_,M_] carries out a numerical
integration of the KdV equation using the procedure of Zabusky & Kruskal.
The input parameter initially determines the initial solution in the procedure;
e.g. -6 Sech^2[x]. The infinitesimals dx and dt are the steps with respect
to the spatial and temporal directions. M fixes the number of steps along
the x-axis."

Begin["`Private`"]

KdVNIntegrate[initial_,dx_,dt_,M_]:=Block[
                    {uPresent, uPast, uFuture, initialh, m, n},
(* --- transform the initial conditions on the grid --- *)
     initialh = initial /. f_[x_] -> f[(m-M/2) dx];
     h = dx;
     k = dt;
(* --- calculate the initial solutions on the grid --- *)
     uPast = Table[initialh, {m,1,M}];
(* --- initialization of the lists containing the grid points
                    uPresent = present       (m)
                    uFuture = future        (m+1)
                    uPast = past           (m-1)            --- *)
     uPresent = uPast;
     uFuture = uPresent;
     ik = 0;
(* --- iteration for the first step --- *)
     Do[
        uPresent[[m]] = uPast[[m]] + 6 k (uPast[[m+1]] +
                   uPast[[m]] + uPast[[m-1]])
                   (uPast[[m+1]] - uPast[[m-1]])/(3 h) -
                   k (uPast[[m+2]] - 2 uPast[[m+1]] + 2 uPast[[m-1]] -
                   uPast[[m-2]])/h^3,
     {m,3,M-2}];
(* --- iterate the time --- *)
     Do[
(* --- iterate the space points --- *)
        Do[
           uFuture[[m]] = uPast[[m]] + 6 k (uPresent[[m+1]] +
                   uPresent[[m]] + uPresent[[m-1]])
                   (uPresent[[m+1]] - uPresent[[m-1]])/(3 h) -
                   k (uPresent[[m+2]] - 2 uPresent[[m+1]] +
                   2 uPresent[[m-1]] -
                   uPresent[[m-2]])/h^3,
        {m,3,M-2}];
(* --- exchange lists --- *)
           uPast = uPresent;
           uPresent = uFuture;
(* --- plot a time step --- *)
        If[Mod[n,40] == 0,
           ik = ik + 1;
(*--- plots are stored in a[1], a[2], ... a[6] ---*)
           a[ik] = ListPlot[uFuture,
```

```
                              AxesLabel->{"x","u"},
                              Prolog->Thickness[0.001],
                              PlotJoined->True,
                              PlotRange->{-15,0.1}]],
           {n,0,320}]
           ]
  End[]
  EndPackage[]
```

5.5 Exercises

1. Using the package **KdVEquation'**, find the type of differential equation for approximating orders $n \geq 3$. Does this approximation change the nonlinearity of the equation? What kinds of effects occur in higher approximations?

2. Change the package **KdVEquation'** so that you can treat arbitrary dispersion relations.

   Caution: Make a copy of the original package first!

3. Examine the motion of the four solitons of the KdV equation. Study the phase gap in the contour plot of the four solitons.

4. Demonstrate that the odd densities of the conservation laws of the KdV equation w_{2n+1} $(n = 0, 1, 2, \ldots)$ are total differentials of the w_{2n}'s.

5. Re-examine the determination of eigenvalues for the anharmonic oscillator. Discuss the link between the eigenvalue problem and the KdV equation.

6. Derive a single soliton solution by using the inverse scattering method for the KdV equation.

7. Examine the numerical solution of the KdV equation for initial conditions which do not satisfy $b(k) = 0$.

8. Change the step intervals in the space and time parameters of the numerical solution procedure for the KdV equation. Examine the accuracy of the numerical integration process. Compare the numerical solution to the analytical solution of the KdV equation.

9. Study the influence of the discretization number M in the numerical integration of the KdV equation.

6

General Relativity

6.1 The orbits in general relativity

From the classical theory of orbital motion we know that a planet in a central force field moves in an ellipse around the center of the planetary system. The orbit of the planet is confined to a plane with fixed orientation. This behavior is in contradiction to the observations made at the turn of the century. From observations of the orbital motion of planets, especially of Mercury, astronomers have discovered that the perihelion of the orbit is rotating. This movement of the perihelion is called perihelion shift. The classical theories of Kepler and Newton do not accurately describe the perihelion shift. The second law of Kepler states that a planet moves in an ellipse around the center of the planetary system. In classical theory, the orbital motion is governed by the conservation of energy and angular momentum. The conservation of angular momentum confines the planet to a plane. Another conserved quantity of Newton's theory is the Lenz vector. The Lenz vector is a vector from the focus to the perihelion that is constant (i.e., in classical theory, the perihelion is at a fixed point in space). In Einstein's general relativity (GR) theory, these assumptions are altered. In GR, the orbits are not closed paths and there exists a perihelion rotation. The actual planetary orbits are rosettes. For these types of orbits the perihelion rotates slowly around the sun. The rotation of the orbit results from two effects [6.1]:

- To calculate the orbit using special relativity, we have to take into account an increase of the mass by

$$m = \frac{m_0}{\sqrt{1 - \frac{v^2}{c^2}}} \qquad (6.1)$$

where m_0 is the rest mass of the planet, c the velocity of light, and v the velocity of the planet in the orbit.

- The central star produces a gravitational field. According to Einstein's theory, this gravitational field is related to an energy density which in turn is directly connected with a mass density. The additional mass density of the field adds a certain amount of field strength to the strength of the sun.

Both effects are relevant in explaining the perihelion shift of a planet. In the following we consider in more detail the second effect [6.1]. The sun of our solar system possesses a much larger mass than the accompanying planets, which means that we can locate the origin of the coordinate system in the sun. Since the orbit is confined to a plane in space (conservation of angular momentum), we can use plane polar coordinates (r, ϕ) to describe the motion of the planets. In GR the distance between two points is not simply given by the radial distance r but is also a function of the radial coordinate. If we denote time by t, we can express the line element ds^2 in space time in the Schwarzschild metric by

$$ds^2 = c^2 \left(1 - \frac{R^s}{r}\right) dt^2 - \frac{dr^2}{1 - R^s/r} - r^2 d\phi^2 \tag{6.2}$$

[6.2] where c denotes the speed of light and $R^s = 2Gm/c^2$ the Schwarzschild radius of the gravitational field. G is the gravitational constant and m the mass of the sun. The Lagrangian of the motion in this metric is given by

$$L = \frac{1}{2}c^2 \left(1 - \frac{R^s}{r}\right) t'^2 - \frac{1}{2}\frac{r'^2}{1 - R^s/r} - \frac{1}{2}r^2\phi'^2, \tag{6.3}$$

where the primes denote differentiation with respect to the line element s. Since GR is a geometrically based theory, the orbits of the theory are derivable by a variational principle. Fermat's principle, which governs the path of a light beam, is an example from optics. In GR the orbits follow from the extremum of the action as determined by the Lagrangian. In close analogy to our considerations in Chapter 1, the equations of motion of GR follow from the Euler–Lagrange equations in the form

$$\frac{d}{ds}\left(\frac{\partial L}{\partial r'}\right) - \frac{\partial L}{\partial r} = 0 \tag{6.4}$$

$$\frac{d}{ds}\left(\frac{\partial L}{\partial \phi'}\right) - \frac{\partial L}{\partial \phi} = 0 \tag{6.5}$$

$$\frac{d}{ds}\left(\frac{\partial L}{\partial t'}\right) - \frac{\partial L}{\partial t} = 0. \tag{6.6}$$

Unlike the classical theory of variation, here we consider time t as a dependent variable. Using (6.3), equations (6.5) and (6.6) lead to angular momentum ℓ and energy conservation.

$$\frac{\partial L}{\partial \phi'} = const. = \ell \tag{6.7}$$

$$\frac{\partial L}{\partial t'} = const. = \mathcal{E}_0 \tag{6.8}$$

or

$$r^2 \phi' \;=\; \ell = \frac{1}{\sqrt{\beta}} \tag{6.9}$$

$$\left(1 - \frac{R^s}{r}\right) t' \;=\; \mathcal{E}_0 = -\frac{k^2}{c^2 \sqrt{\beta}}, \tag{6.10}$$

where k and β are appropriate constants for the following considerations. Using the conserved quantities in the expression of the line element (6.2), we get

$$\frac{dr^2}{1 - R^s/r} = \left\{ -\beta r^4 + \frac{k^2 r^4}{c^2 (1 - R^s/r)} - r^2 \right\} d\phi^2. \tag{6.11}$$

Substituting $u = 1/r$ simplifies the equation of the orbit to

$$\left(\frac{du}{d\phi}\right)^2 = \frac{k^2}{c^2} - (1 - R^s u)(\beta + u^2). \tag{6.12}$$

This exact equation is usually solved by using the perturbation theory which approximates the solution for a certain range [6.3, 6.4]. At the end of this section we demonstrate how the equation (6.12) can be solved exactly using *Mathematica*. Since the equation consists of a polynomial of third order in u, the solution of (6.12) is expressible by elliptic functions. To see how this occurs we make the transformation

$$u = \frac{4}{R^s} U + \frac{1}{3R^s} \tag{6.13}$$

and substitute it into equation (6.12). The resulting differential equation is the defining equation for the Weierstrass function $\mathcal{P}(z)$

$$\left(\frac{dU}{d\phi}\right)^2 = 4U^3 - g_2 U - g_3 \tag{6.14}$$

where

$$g_2 \;=\; \frac{1}{12} - \frac{(R^s)^2 \beta}{4} \tag{6.15}$$

$$g_3 \;=\; \frac{1}{216} + \frac{(R^s)^2 \beta}{24} - \frac{(R^s)^2 k^2}{16c^2}. \tag{6.16}$$

The solution of U is thus

$$U = \mathcal{P}(\phi + C; g_2, g_3), \tag{6.17}$$

where C denotes the integration constant. The orbits are now represented by the coordinates r and ϕ

$$r(\phi) = \frac{3R^s}{1 + 12\mathcal{P}(\phi + C; g_2, g_3)} \tag{6.18}$$

6.1.1 Quasi elliptic orbits

If g_2 and g_3 are real and the discriminant $\Delta = g_2^3 - 27g_3^2 > 0$ we find three real roots of the characteristic polynomial $4x^3 - g_2 x - g_3 = 0$ which we call e_1, e_2, and e_3. The roots of the characteristic polynomial can be arranged in the order $e_2 < e_3 < e_1$. Using the roots and the expressions g_1 and g_2, we can express the periods ω_1 and ω_2 of the Weierstrass function by

$$\omega_1 = \int_{e_1}^{\infty} \frac{dx}{\sqrt{4x^3 - g_2 x - g_3}} \tag{6.19}$$

and

$$\omega_2 = i \int_{-\infty}^{e_2} \frac{dx}{\sqrt{4x^3 - g_2 x - g_3}}. \tag{6.20}$$

The first period ω_1 is a real and the second period ω_2 is an imaginary number. ω_2 is the period of the angle ϕ. If we introduce a third frequency ω_3, the equation of the orbit (6.18) is expressible in the form

$$r(\phi) = \frac{3R^s}{1 + 12\mathcal{P}(\phi - \omega_3; g_2, g_3)}. \tag{6.21}$$

By introducing ω_3, we are able to suppress the singularity of the Weierstrass function at $z = 0$. The correct specification of the orbit is made by the choice of the locations of the perihelion and the aphelion. Choosing the coordinate system so that the perihelion is reached at $\phi = 0$, we get from equation (6.21)

$$r(0) = \frac{3R^s}{1 + 12\mathcal{P}(-\omega_3)} = \frac{3R^s}{1 + 12e_3} \tag{6.22}$$

$$\frac{dr^{-1}}{d\phi} = 0 \tag{6.23}$$

and

$$\frac{d^2 r^{-1}}{d\phi^2} < 0. \tag{6.24}$$

Once the planet has approached the aphelion, it has traced one half of the total orbit. This point of the orbit is characterized by the angle $\phi = \omega_1$. The radial coordinate at this point is expressed by

$$r(\omega_1) = \frac{3R^s}{1 + 12\mathcal{P}(\omega_1 - \omega_3)} = \frac{3R^s}{1 + 12\mathcal{P}(\omega_2)} = \frac{3R^s}{1 + 12e_2} \tag{6.25}$$

$$\frac{dr^{-1}}{d\phi} = 0 \tag{6.26}$$

and

$$\frac{d^2 r^{-1}}{d\phi^2} > 0. \tag{6.27}$$

The relations (6.25) and (6.27) are correct if the condition $\frac{1}{12} + e_2 > 0$ is satisfied. This condition is equivalent to the relation $c^2\beta > k^2$ w, relating the parameters of the Weierstrass function to the physical parameters of the path. The radial coordinate of the orbit varies between the limits of the perihelion and the aphelion measured from the origin of the coordinate system. The two extreme values of the orbit are

$$r_P = \frac{3R^s}{1 + 12e_3} \tag{6.28}$$

$$r_A = \frac{3R^s}{1 + 12e_2}. \tag{6.29}$$

The planet is thus confined between two circles with radii r_P and r_A. The path itself is an open orbit in the form of a rosette (see figure 6.1, where only the path is shown). The orbit in figure 6.1 is similar to the classical orbit of Kepler's theory. Unlike the classical orbit, the GR shows shifts of the perihelion and the aphelion. From the classical theory of planet motion, we know that the difference of phase between two complete rotations is given by $\phi = 2\pi$. Within GR the difference in the angle is exactly $2\omega_1$. The shift in the perihelion is thus determined by

$$\Delta\phi^P = 2(\pi - \omega_1). \tag{6.30}$$

The perihelion shift in the solar system is very small and its experimental observation is very difficult. However, the calculation of equation (6.30) needs to be precise in order to determine the exact numerical value of the perihelion shift. To calculate the shift using the Weierstrass function, we need an absolute accuracy of 10^{-8} in the values for $\mathcal{P}(z)$. In a graphical representation of the Mercury orbit, for example, the shift is invisible. The observed and calculated shift for Mercury is $43.1''$ for 415 cycles (approximately one century).

The perihelion and the aphelion are determined by relation (6.28). The locations of the perihelion and the aphelion are usually given by the classical parameters, the major semi axis a and eccentricity e. If we combine both parameters of GR and classical theory, we get the relations for r_P and r_A

$$r_P = \frac{p}{1 + e} \tag{6.31}$$

$$r_A = \frac{p}{1 - e} \tag{6.32}$$

where $p = b^2/a$ and $e = \sqrt{a^2 - b^2}/a$. Having determined the extreme points of the orbit, we know the roots of the Weierstrass function $\mathcal{P}$, e_2 and e_3, from relation (6.28). The roots are given by

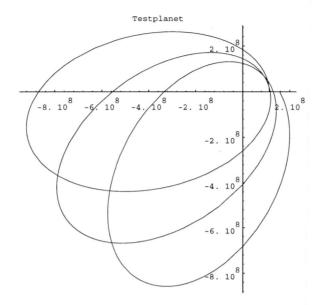

Figure 6.1.   Perihelion shift for a system of planets with $m = 5.6369 \cdot 10^{33}\,kg$, $a = 5.2325 \cdot 10^{8}\,m$ and eccentricity $e = 0.61713$. The numeric value of the perihelion shift is calculated to be $\Delta\phi^P = 90122.8''$.

$$e_2 = -\frac{1}{12}\left\{1 - \frac{3R^s}{r_A}\right\} \tag{6.33}$$

$$e_3 = -\frac{1}{12}\left\{1 - \frac{3R^s}{r_P}\right\}. \tag{6.34}$$

In terms of the orbit parameters, we find

$$e_2 = -\frac{1}{12}\left\{1 - \frac{3R^s a}{b^2}\left(1 - \frac{\sqrt{a^2 - b^2}}{a}\right)\right\} \tag{6.35}$$

$$e_3 = -\frac{1}{12}\left\{1 - \frac{3R^s a}{b^2}\left(1 + \frac{\sqrt{a^2 - b^2}}{a}\right)\right\}. \tag{6.36}$$

The roots of the $\mathcal{P}$-function have to satisfy relations

$$0 = e_1 + e_2 + e_3 \tag{6.37}$$

$$g_2 = 2(e_1^2 + e_2^2 + e_3^2) \tag{6.38}$$

$$g_3 = 4e_1 e_2 e_3. \tag{6.39}$$

Here the root e_1 becomes

$$e_1 = \frac{1}{6}\left\{1 - \frac{3aR^s}{b^2}\right\}. \tag{6.40}$$

The quantities g_2 and g_3 are determined by the expressions (6.15) and (6.16) and satisfy relations (6.38-6.39). We are now able to determine the energy $\mathcal{E}_0$ and the angular momentum ℓ from the orbital parameters from (6.9–6.10). The angular momentum and the energy can be represented by

$$\ell = \frac{R^s}{2(\frac{1}{12} - g_2)^{1/2}} \tag{6.41}$$

$$\mathcal{E}_0 = -\frac{2}{c}\left\{\frac{\frac{1}{54} - \frac{1}{6}g_2 - g_3}{\frac{1}{12} - g_2}\right\}^{1/2}. \tag{6.42}$$

One problem with using the exact solution theory is the determination of the angles ω_1 and ω_2 when calculating the perihelion shift with *Mathematica*. As mentioned before, we need a high degree of accuracy in our calculation to find the right value for $\Delta\phi$. If we do the calculations by simply integrating equations (6.19) and (6.20), we have a singularity at one of the end points of the integration interval. Since we have no convergent representation of the integral, the results are very crude. However, we know from the theory of the Weierstrass functions that the periods are expressible by complete elliptic integrals of the first kind. Using the properties of the elliptic integrals, we can overcome the inaccurate numerical integration of *Mathematica*.

$$\omega_1 = \frac{K(m)}{\sqrt{e_1 - e_2}} \tag{6.43}$$

$$\omega_2 = i\frac{K'(m)}{\sqrt{e_1 - e_2}} = i\frac{K(1-m)}{\sqrt{e_1 - e_2}}, \tag{6.44}$$

where the module m is given by $m = (e_3 - e_2)/(e_1 - e_2)$, the roots of the Weierstrass function.

The above considerations are collected in the following *Mathematica* package.

──────────────── File periheli.m ────────────────

```
BeginPackage["PerihelionShift`"]

Clear[e1,e2,e3,g2,g3,omega1,omega2,Orbit,orbit,Energy,AngularMomentum,
    PerihelionShift,Planets,DOOrbit,Schwarzschild];

Planets::usage = "Planets[planet_String] creates a list of data for planets and
planetoids stored in the data base of the package PerihelionShift. The data
base contains the names of the planets, their major axes, their eccentricity
and the mass of the central planet. Planets['List'] creates a list of the
planets in the data base. Planets['name'] delivers the data of the planet
given in the argument."

orbit::usage ="orbit[phiend_,minorAxes_,majorAxes_,mass_]
creates a graphical representation of the perihelion shift if the major and
minor axes and the mass are given."
```

```
Orbit::usage ="Orbit[planet_String] creates a graphical representation of the
perihelion shift for the planets contained in the data base."

PerihelionShift::usage ="PerihelionShift[minorAxes_,majorAxes_,mass_]
Calculates the numerical value of the perihelion shift."

AngularMomentum::usage = "AngularMomentum[minorAxes_,majorAxes_,mass_]
calculates the angular momentum of a planet."

Energy::usage ="Energy[minorAxes_,majorAxes_,mass_]
calculates the energy of a planet."

DOOrbit::usage ="DOOrbit[planet_String,phiend_,options___] plots the orbit
in the case of vanishing determinants (see text)."


Begin["'Private'"]

(* --- data bases of several planets --- *)

data = {{"Mercury",0.5791 10^(11),0.2056,MassOfTheSun},
        {"Venus",1.0821 10^(11),0.0068,MassOfTheSun},
        {"Earth",1.4967 10^(11),0.0167,MassOfTheSun},
        {"Icarus",1.61 10^(11),0.827,MassOfTheSun},
        {"Mars",2.2279 10^(11),0.093,MassOfTheSun},
        {"Ceres",4.136 10^(11),0.076,MassOfTheSun},
        {"Jupiter",7.78 10^(11),0.048,MassOfTheSun},
        {"Saturn",14.27 10^(11),0.056,MassOfTheSun},
        {"Uranus",28.70 10^(11),0.047,MassOfTheSun},
        {"Neptune",44.96 10^(11),0.009,MassOfTheSun},
        {"Pluto",59.10 10^(11),0.25,MassOfTheSun},
        {"PSR1916",7.0204020286 10^(8),0.6171313,2.82837 MassOfTheSun},
        {"TestPlanet",5.2327 10^(8),0.6171313,2828.37 MassOfTheSun}}

(* --- information on the planets --- *)

Planets[planet_String]:=Block[{gh,kh,ma},
 MassOfTheSun = 1.993 10^(30);
 If[planet == "List",
     Do[Print["   ",data[[k,1]]],{k,1,Length[data]}],
     gh = 0;
     kh = 0;
     ma = 0;
     Do[
            If[planet == data[[k,1]],
            Print["----------------------------------------"];
            Print["   ",data[[k,1]]];
            Print["----------------------------------------"];
            Planet = data[[k,1]];
            gh = data[[k,2]];
            kh = N[data[[k,2]] Sqrt[1-data[[k,3]]]];
            ma = data[[k,4]];
            Print["Mass           = ",ma];
            Print["minor axes     = ",kh];
            Print["major axes     = ",gh];
            Print[" "];
            Print["eccentricity   = ",data[[k,3]]];
```

```
            Print["-----------------------------------------"],
            gh = gh;
            kh = kh;
            ma = ma],
        {k,1,Length[data]}];
        MajorAxes = gh;
        MinorAxes = kh;
        Mass = ma;
        If[gh != 0,
        PerihelionShift[kh,gh,ma],0]]]

(* --- Schwarzschild radius --- *)

SchwarzSchild[mass_]:=
     Block[{Gravitation, SpeedOfLight},
     Gravitation = 6.6732 10^(-11);
     SpeedOfLight = 2.9979250 10^8;
     2 Gravitation mass/SpeedOfLight^2
     ]

(* --- roots of the characteristic polynomial --- *)

e2[minorAxes_,majorAxes_,mass_]:=
     Block[{Schwarzschild,eh},
     Schwarzschild = SchwarzSchild[mass];
     eh = -(1-3 majorAxes Schwarzschild/minorAxes^2
          (1-Sqrt[majorAxes^2-minorAxes^2]/
          majorAxes))/12];

e3[minorAxes_,majorAxes_,mass_]:=
     Block[{Schwarzschild,eh},
     Schwarzschild = SchwarzSchild[mass];
     eh = -(1-3 majorAxes Schwarzschild/minorAxes^2
          (1+Sqrt[majorAxes^2-minorAxes^2]/
          majorAxes))/12];

e1[minorAxes_,majorAxes_,mass_]:=Block[{},
          -( e3[minorAxes,majorAxes,mass] +
             e2[minorAxes,majorAxes,mass])];

(* --- g2 and g3 of the Weierstrass function --- *)

g2[minorAxes_,majorAxes_,mass_]:=Block[{},
          2 (e1[minorAxes,majorAxes,mass]^2 +
             e2[minorAxes,majorAxes,mass]^2 +
             e3[minorAxes,majorAxes,mass]^2)];

g3[minorAxes_,majorAxes_,mass_]:=Block[{},
          4 e1[minorAxes,majorAxes,mass]
            e2[minorAxes,majorAxes,mass]
            e3[minorAxes,majorAxes,mass]];

(* --- frequencies of the Weierstrass function --- *)

omega1[minorAxes_,majorAxes_,mass_]:=
     Block[{integrand,x,om1,e11,e21,e31,module},
     integrand = 4 x^3 - g2[minorAxes,majorAxes,mass] x -
```

```
                g3[minorAxes,majorAxes,mass];
        integrand = 1/Sqrt[integrand];
        e11 = e1[minorAxes,majorAxes,mass];
        e21 = e2[minorAxes,majorAxes,mass];
        e31 = e3[minorAxes,majorAxes,mass];
        module = (e31-e21)/(e11-e21);
        om1 = EllipticK[module]/Sqrt[e11-e21]
          ];

omega2[minorAxes_,majorAxes_,mass_]:=
        Block[{integrand,x,om2,e11,e21,e31,module},
        integrand = Abs[4 x^3 -
                    g2[minorAxes,majorAxes,mass] x -
                    g3[minorAxes,majorAxes,mass]];
        integrand = 1/Sqrt[integrand];
        e11 = e1[minorAxes,majorAxes,mass];
        e21 = e2[minorAxes,majorAxes,mass];
        e31 = e3[minorAxes,majorAxes,mass];
        module = (e31-e21)/(e11-e21);
        module = 1 - module;
        om2 = I EllipticK[module]/Sqrt[e11-e21]
          ];

(* --- creates the orbit from the orbit parameters --- *)

orbit[phiend_,minorAxes_,majorAxes_,mass_]:=
        Block[{Schwarzschild,bh,omega3,l2,l3,l4,l5,phi},
        Schwarzschild = SchwarzSchild[mass];
        om1 = omega1[minorAxes,majorAxes,mass];
        om2 = omega2[minorAxes,majorAxes,mass];
        omega3 =  om1 + om2;
        l2 = g2[minorAxes,majorAxes,mass];
        l3 = g3[minorAxes,majorAxes,mass];
        l4 = Chop[WeierstrassP[phi-omega3,l2,l3]];
        l5 = 1 + l2 l4;
        bh = 3 Schwarzschild/l5;
        ParametricPlot[{Cos[phi] bh,Sin[phi] bh},{phi,0,phiend},
                    PlotRange->All,
                    Prolog->Thickness[0.001],
                    PlotLabel->planet]]

(* --- creates the orbit with the data base --- *)

Orbit[planet_String]:=Block[{},
        Planets[planet];
        orbit[6Pi,MinorAxes,MajorAxes,Mass]]

(* --- numerical value of the perihelion shift --- *)

PerihelionShift[minorAxes_,majorAxes_,mass_]:=
        Block[{ph,ph1},
        ph = N[2 (omega1[minorAxes,majorAxes,mass]-Pi),16];
        ph1 = ph  2.06264806245 10^5;
        Print[" "];
        Print[" Perihelion shift = ",ph1," arcs"];
        ph]
```

```
(* --- constants of motion --- *)

AngularMomentum[minorAxes_,majorAxes_,mass_]:=
     Block[{Schwarzschild,l1},
     Schwarzschild = SchwarzSchild[mass];
     l1 = g2[minorAxes,majorAxes,mass];
     l1 = Schwarzschild/(2 (1/12 - l1))]

Energy[minorAxes_,majorAxes_,mass_]:=
     Block[{Schwarzschild,energy,l2,l3},
     Schwarzschild = SchwarzSchild[mass];
     l2= g2[minorAxes,majorAxes,mass];
     l3= g3[minorAxes,majorAxes,mass];
     energy = -2 Sqrt[(1/54 -l2/6 -l3)/(1/12 - l2)]/SpeedOfLight]

(* --- asymptitic orbits --- *)

DOOrbit[planet_String,phiend_,options___]:=
     Block[{Schwarzschild,e0,n2,phi},
     Planets[planet];
     Schwarzschild = SchwarzSchild[mass];
     e0 = 1/24 - Schwarzschild/(4 MajorAxes);
     n2 = 3 e0;
     bh1 = 4/Schwarzschild    (1/12 + n2/3 - n2/Cosh[Sqrt[n2] phi]^2);
     bh1 = 1/bh1;
     ParametricPlot[{Cos[phi] bh1,Sin[phi] bh1},{phi,-phiend,phiend},
                    options]]
End[]
EndPackage[]
```

6.1.2 Asymptotic circles

In this section we discuss a limiting case of GR orbits which is closely related to the classical orbits of the Kepler theory. We assume that the constants k and β are such that the discriminant Δ vanishes. For this case, two of the roots e_1, e_2, e_3 are the same. If we denote the common root by e, the remaining root takes the value $-2e$. For $e > 0$, the solution of the orbit equation (6.18) is

$$r(\phi) = \frac{3R^s \cosh(n\phi)}{1 - 8n^2}, \tag{6.45}$$

where $n^2 = 3e$. This solution results in an apogee with $\phi = 0$; provided that $8n^2 < 1$. This restriction is equivalent to the condition $(R^s)^2 \beta > \frac{1}{4}$.

If ϕ increases, the orbit of the planet spirals down to a circle of asymptotic radius

$$r = \frac{3R^s}{1 + 4n^2}. \tag{6.46}$$

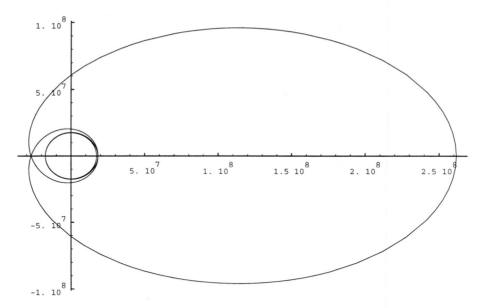

Figure 6.2.   Orbit for a test planet with $\Delta = 0$.

This radius is smaller than the initial distance of the planet from the center of the planetary system (see figure 6.2). If we choose n so that the relation $0 < n^2 < \frac{1}{8}$ is satisfied, the radius of the asymptotic circle lies between the limits $3R^s$ and $2R^s$. The orbit for such cases is obtained by function **D0Orbit[ ]** defined in the package **PerihelionShift'**.

6.2   Light bending in the gravitational field

GR predicts that a light ray is bent in a gravitational field. The corresponding equation of motion follows from the null geodesic condition $ds^2 = 0$ [6.2]. We discuss the bending of a light ray in the Schwarzschild metric. The equation of motion is given by

$$\ddot{u} + u - \frac{3}{2}R^s u^2 = 0 \tag{6.47}$$

where $u = 1/r$ and $R^s = 2Gm/c^2$ is the Schwarzschild radius of the mass m. G denotes the gravitational constant and c the speed of light. Multiplying equation (6.47) by $\dot{u} = du/d\phi$ and integrating it with respect to parameter s we get

$$\frac{1}{2}\dot{u}^2 + \frac{1}{2}u^2 - \frac{R^s}{2}u^3 = E = \frac{k^2}{c^2}. \tag{6.48}$$

where E and k, energy and scaled energy, are appropriately chosen constants. The substitution $u = 4U/R^s + 1/(3R^s)$ transforms equation (6.48) to a standard form of differential equations defining the Weierstrass function

$$\left(\frac{dU}{d\phi}\right)^2 = 4U^3 - g_2U - g_3 \tag{6.49}$$

with

$$\begin{aligned}
g_2 &= \frac{1}{12} \\
g_3 &= \frac{1}{216} - \frac{(R^s)^2 k^2}{16c^2}.
\end{aligned} \tag{6.50}$$

The solution for the variable U is given by

$$U(\phi) = \mathcal{P}(\phi + C; g_2, g_3). \tag{6.51}$$

The path of the light ray $r(\phi)$ is

$$r(\phi) = \frac{3R^s}{1 + 12\mathcal{P}(\phi + C; g_2, g_3)}. \tag{6.52}$$

The geometrical locations of the planet and the light ray are given in figure 6.3. Figure 6.3 shows that the light ray has a distance R from the planet if the angle $\phi = 0$.

When $\phi = \phi_1$, the radius (6.52) is infinite. The deviation or bending of the light ray $\delta\phi$ is determined by the relation

$$\delta\phi = 2\phi_1 - \pi \tag{6.53}$$

(see figure 6.3). Since the Schwarzschild radius R^s and the constant k^2/c are greater than zero it follows that the discriminant $\Delta = g_2^3 - 27g_3^2 > 0$.

The equation $r(\phi = 0) = R$ gives us the first expression for the determination of the roots e_1, e_2, e_3 of the characteristic polynomial $4t^3 - g_2 t - g_3 = 0$. If we set $\phi = 0$, it follows from equation (6.52)

$$r(\phi = 0) = R = \frac{3R^s}{1 + 12\mathcal{P}(C; g_2, g_3)}. \tag{6.54}$$

If we choose the integration constant as the imaginary period of the Weierstrass function $C = -\omega_2$, we get from the condition $\mathcal{P}(-\omega_2) = e_2$ the relation

$$R = \frac{3R^s}{1 + 12e_2} \tag{6.55}$$

and thus $e_2 = -(1 - 3R^s/R)/12$. Since g_2 is fixed to $1/12$ in the light bending problem, the remaining two roots e_1 and e_3 satisfy

$$\begin{aligned}
g_2 &= 2(e_1^2 + e_2^2 + e_3^2) = \frac{1}{12} \tag{6.56} \\
0 &= e_1 + e_2 + e_3. \tag{6.57}
\end{aligned}$$

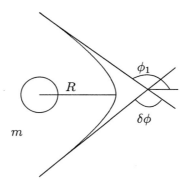

Figure 6.3. Geometry of light bending in the neighborhood of a mass m. The deviation angle ϕ_1 follows from the relations $\phi_2 = \pi - \phi_1$ and $\delta\phi = \pi - 2\phi_2 = 2\phi_1 - \pi$.

We find, by eliminating $e_3 = -(e_2 + e_1)$, in equation (6.56), the relation

$$e_1^2 + e_2 e_1 + e_2^2 - \frac{1}{4}g_2 = 0 \tag{6.58}$$

has the solution

$$e_1 = -\frac{1}{2}e_2 \pm \frac{\sqrt{3}}{12}\sqrt{1 - 36e_2^2}. \tag{6.59}$$

From (6.57) we can derive the solution for e_1 to be

$$e_3 = -(e_2 + e_1) = -\left\{\frac{1}{2}e_2 \pm \frac{\sqrt{3}}{12}\sqrt{1 - 36e_2^2}\right\}. \tag{6.60}$$

The remaining problem is to find the angle of inclination, i.e., the angle ϕ_1 for which the radius tends to infinity. We can express this condition by

$$r(\phi = \phi_1) = \infty = \frac{3R^s}{1 + 12\mathcal{P}(\phi_1 - \omega_2; g_2 g_3)}. \tag{6.61}$$

Equation (6.61) is satisfied if

$$\mathcal{P}(\phi_1 - \omega_2; g_2, g_3) + \frac{1}{12} = 0. \tag{6.62}$$

The frequency ω_2 is derived from the roots e_1, e_2, e_3 and satisfies the relations

$$\omega_2 = \omega + \omega' \tag{6.63}$$
$$\omega_1 = \omega \qquad \text{real} \tag{6.64}$$
$$\omega_3 = \omega' \qquad \text{imaginary.} \tag{6.65}$$

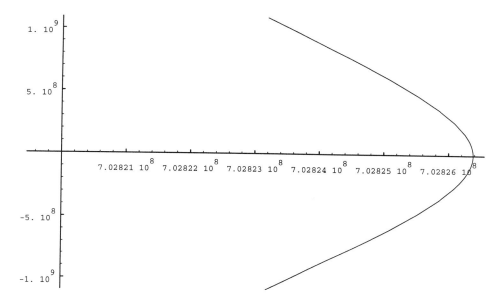

Figure 6.4.   Path of a light ray in the neighborhood of the sun.

In addition, there are two relations for the frequencies ω and ω'

$$\omega = \frac{K(m)}{\sqrt{e_1 - e_3}} \quad \text{and} \quad \omega' = i\frac{K(1-m)}{\sqrt{e_1 - e_3}}, \tag{6.66}$$

where the modulus $m = (e_2 - e_3)/(e_1 - e_3)$. Equation (6.62) is only solvable numerically and provides us with the limiting angle ϕ_1. The angle determines the asymptotic direction of the light ray. An example of the bending of a light ray near the surface of the sun is shown in figure 6.4. The graphical representation of the light bending is created using **Orbit[]**, a function of the package **LightBending'**. The function **Deviation[]**, which is also contained in this package, allows the numerical calculation of the bending angle. The arguments of **Deviation[]** are the mass of the planet and the closest approach of the light ray.

―――――――――――― File lightben.m ――――――――――――

```
BeginPackage["LightBending'"]


Clear[e1, e2, e3, g2, g3, omega1, omega2, Orbit, Deviation];

Deviation::usage = "Deviation[radius_,mass_] calculates the numerical value
of the light bending in a gravitational field of a planet with mass M in a
distance radius of the center."

Orbit::usage = "Orbit[radius_,mass_] plots the orbit of a light beam near
a mass in the distance radius. The calculation is done in Schwarzschild
```

```
metric."

MassOfTheSun::usage
RadiusOfTheSun::usage

Begin["'Private'"]

(* --- mass and radius of the sun --- *)

MassOfTheSun = 1.993 10^(30);
RadiusOfTheSun = 7 10^8;

(* --- Schwarzschild radius --- *)

SchwarzSchild[mass_]:=
     Block[{Gravitation,SpeedOfLight},
     Gravitation = 6.6732 10^(-11);
     SpeedOfLight = 2.9979250 10^8;
     Schwarzschild = 2 Gravitation mass/SpeedOfLight^2
     ]

(* --- roots of the characteristic polynomial --- *)

e1[radius_,mass_]:=
     Block[{eh,e31},
     e21 = e2[radius,mass];
     eh = N[-1/2 e21 + Sqrt[3] Sqrt[1-36 e21^2]/12]]

e2[radius_,mass_]:=
     Block[{Schwarzschild,eh},
     Schwarzschild = SchwarzSchild[mass];
     eh = -1/12 (1 - 3 Schwarzschild/radius)
          ]

e3[radius_,mass_]:=
     Block[{eh},
     eh = N[-(e2[radius,mass] + e1[radius,mass])]]

(* --- frequencies of the Weierstrass function --- *)

omega1[radius_,mass_]:=
     Block[{om1,e11,e21,e31,modulus},
     e11 = e1[radius,mass];
     e21 = e2[radius,mass];
     e31 = e3[radius,mass];
     modulus = (e21-e31)/(e11-e31);
     om1 = EllipticK[modulus]/Sqrt[e11-e31]
       ];

omega2[radius_,mass_]:=
     Block[{om2,e11,e21,e31,modulus},
     e11 = e1[radius,mass];
     e21 = e2[radius,mass];
     e31 = e3[radius,mass];
     modulus = (e21-e31)/(e11-e31);
     modulus = 1 - modulus;
     om2 = I EllipticK[modulus]/Sqrt[e11-e31]
```

```
        ];

  (* --- g2 and g3 of the Weierstrass function --- *)

  g2[radius_,mass_]:=Block[{},N[1/12]]

  g3[radius_,mass_]:=Block[{},
        4 e1[radius,mass] e2[radius,mass] e3[radius,mass]]

  (* --- creates the path of the light beam --- *)

  Orbit[radius_,mass_]:=
        Block[{Schwarzschild,bh,l2,l3,l4,l5,phi,phia,deltaphi,
                erg,omega3},
        Schwarzschild = SchwarzSchild[mass];
        om1 = omega1[radius,mass];
        om2 = omega2[radius,mass];
        omega3 = om1 + om2;
        l2 = g2[radius,mass];
        l3 = g3[radius,mass];
        l4 = WeierstrassP[phi-omega3,l2,l3]+1/12;
        erg = FindRoot[l4==0,{phi,Pi/2}];
        phia = phi /. erg;
        phia = Re[phia];
        l4 = Re[WeierstrassP[phi-omega3,l2,l3]];
        l5 = 1 + l2 l4;
        bh = 3 Schwarzschild/l5;
        ParametricPlot[{Cos[phi] bh,Sin[phi] bh},
                        {phi,-phia 0.9,phia 0.9},
                        Prolog->Thickness[0.001]]
        ]

  (* --- determination of the deviation angle --- *)

  Deviation[radius_,mass_]:=
        Block[{Schwarzschild,om1,om2,omega3,l2,l3,l4,phi,
                deltaphi,dphi,phia,erg},
        Schwarzschild = SchwarzSchild[mass];
        om1 = omega1[radius,mass];
        om2 = omega2[radius,mass];
        omega3 = om1+om2;
        l2 = g2[radius,mass];
        l3 = g3[radius,mass];
        l4 = WeierstrassP[phi-omega3,l2,l3]+1/12;
        erg = FindRoot[l4==0,{phi,Pi/2}];
        phia = phi /. erg;
        phia = Re[phia];
        deltaphi = N[2 phia-Pi,16];
  (* --- the factor 2.06264806245 10^5 converts radian to arcsecond --- *)
        dphi = deltaphi   2.06264806245 10^5;
        Print[" "];
        Print[" Deviation = ",dphi," arcs"];
        deltaphi]
End[]
EndPackage[]
```

6.3 Einstein's field equations (vacuum case)

Einstein's theory of gravitation can be described by Riemannian geometry. In Riemannian geometry, space is characterized by its metric. The metric is normally represented by its line element ds^2 or equivalently by the metric tensor which can be read from the line element. The metric tensor allows the calculation of the scalar product of two vectors as well as the equations of motion. Einstein's field equations are the central equations of GR and describe the motion of a particle in space time. Since GR is primarily based on geometry we have to consider the related metric of the space in addition to the physical problem. For our considerations we assume that the independent variables in the space are given by

```
IndepVar={t,x,y,z}

{t, x, y, z}
```

The coordinates are used in the determination of the metric tensor. The function **metric**[] calculates the coefficients of the metric tensor from a given line element. **metric**[] takes the line element ds^2 and a list of coordinates as input variables. The result is the symmetric metric tensor of the underlying space.

```
metric[lineelement_,independentvars_List]:=Block[
{lenindependent,differentials,diffmatrix,
metricform,varmetric,gh,sum,equation,rule,
varhelp,zeros,zerorule},

(* --- determine the number of independent variables ---*)
    lenindependent = Length[independentvars];

(* --- create the differentials corresponding to dx,dt .... --- *)
    differentials = Map[Dt,independentvars];

(* --- a matrix of differential products --- *)
    diffmatrix = Outer[Times,differentials,
                        differentials];

(* --- the general metricform --- *)
    metricform = Array[gh,{lenindependent,
                            lenindependent}];
    varmetric = Variables[metricform];

(* --- built a system of equations to determine
        the elements of the metric            ---*)
    If[Length[metricform] == Length[diffmatrix],
        sum = 0;
```

```
      Do[
        Do[
           sum = sum +
               metricform[[i,j]] diffmatrix[[i,j]],
          {j,1,lenindependent}],
        {i,1,lenindependent}],
      sum = 0
      ];

(* --- construct the metric tensor --- *)
      If[ sum === 0,
         Return[sum],
         sum = sum - lineelement;
         equation = CoefficientList[sum,
                              differentials]==0;
         rule = Solve[equation,varmetric];
         metricform = metricform /. rule;
         varmetric = Variables[metricform];

(* --- replace the nonzero elements --- *)
           varhelp = {};
           Do[
             If[Not[FreeQ[varmetric[[i]],gh]],
                AppendTo[varhelp,varmetric[[i]] ]
              ],
           {i,1,Length[varmetric]}];
           zeros = Table[0,{Length[varhelp]}];
           SubstRule[x_,y_]:=x->y;
           zerorule = Thread[SubstRule[varhelp,zeros]];
           metricform = Flatten[metricform /.
                              zerorule,1];

(* --- make the metricform symmetric --- *)
           metricform = Expand[(metricform +
                         Transpose[metricform])/2]
         ];
      metricform
];
```

6.3.1 Examples for metric tensors

As a first example we consider a simple metric of a hypothetical two-dimensional space in x and t coordinates. The *Mathematica* symbol **Dt[x]** expresses the differential dx in line elements.

```
metric[x Dt[x]^2 + t x Dt[t]^2,{x,t}]//MatrixForm

x      0

0      t x
```

The result is a 2×2 matrix containing the coefficients of the line element. A simple three-dimensional example is the Euclidean space with the well known Cartesian metric. The corresponding line element is $ds^2 = dx^2 + dy^2 + dz^2$.

In *Mathematica* we get the metric by

```
metric[Dt[x]^2+Dt[y]^2+Dt[z]^2,{x,y,z}]

{{1, 0, 0}, {0, 1, 0}, {0, 0, 1}}
```

which is the expected result for the metric tensor. We see that **metric[]** extracts the metric tensor from the line element. The information contained in the metric tensor is of some importance in the derivation of the field equations.

The line element or the metric tensor for Euclidean space changes its form if we use a different coordinate system, e.g., the transformation from Cartesian coordinates to spherical coordinates. In spherical coordinates the metric tensor is given by

```
metric[Dt[r]^2 + r^2 Dt[th]^2 + r^2 Sin[th] Dt[phi]^2,
       {r,th,phi}] //MatrixForm

1           0           0


            2
0           r           0


                        2
0           0           r  Sin[th]
```

where r is the radius, and *phi* and *th* are the spherical polar angles.

A nontrivial example in three dimensions characterizing a curved space is given by the line element $ds^2 = dr^2 + r^2 d\phi^2 + dz^2$ in cylindrical coordinates r, ϕ and z. The corresponding metric tensor is

```
metric[Dt[r]^2+r^2 Dt[phi]^2+Dt[z]^2,
       {r,phi,z}]//MatrixForm

1   0   0


    2
0   r   0


0   0   1
```

In four dimensions, three space dimensions and one time coordinate, the space corresponding to Euclidean space in three dimensions is the Minkowski space. Euclidean space with Cartesian coordinates x, y and z is extended by an additional time dimension t. Note the sign difference when distinguishing between the time coordinate and the space time dimensions. The line element in x, y, z and t is given by $ds^2 = dt^2 - dx^2 - dy^2 - dz^2$ (speed of light equals unity, $c = 1$). The corresponding metric tensor of Minkowski space reads

```
metric[Dt[t]^2 - Dt[x]^2 - Dt[y]^2 -
              Dt[z]^2,{t,x,y,z}]

 {{1, 0, 0, 0}, {0, -1, 0, 0}, {0, 0, -1, 0}, {0, 0, 0, -1}}
```

The Minkowski space is a trivial solution of Einstein's field equations for the vacuum case. A time–independent solution of the field equations with spherical symmetry is the famous Schwarzschild solution. The line element ds^2 in the coordinates t, r, th and phi is

```
ds2 = (1-2 m/r) Dt[t]^2 - (1- 2 m/r)^(-1) Dt[r]^2 -
  r^2 (Dt[th]^2 + Sin[th]^2 Dt[phi]^2)

          2
    Dt[r]               2 m        2
  -(-------) + (1 - ---) Dt[t]   -
      2 m             r
   1 - ---
        r

     2        2         2        2
    r  (Dt[th]  + Dt[phi]  Sin[th] )
```

The corresponding metric is

```
erg = metric[ds2,{t,r,th,phi}]

         2 m                    r                      2
  {{1 - ---, 0, 0, 0}, {0, -------, 0, 0}, {0, 0, -r , 0},
          r                  2 m - r

             2       2
    {0, 0, 0, -(r  Sin[th] )}}
```

This representation of the line element is a spherically symmetric solution of the vacuum field equations. The time–like coordinate t can be interpreted as the world time. The coordinates *th* and *phi* can be identified as the usual angles in spherical coordinates.

The above line element ds^2 resembles the line element in Euclidean space. In the following example the radial coordinate r is transformed so that we can write the line element in an isotropic form $ds^2 = G[\rho]dt^2 - F[\rho](d\rho^2 + \rho^2 dth^2 + \rho^2 sin[th]^2 d\phi^2)$. The transformation reads $r = \rho(1 + m/(2\rho))^2$. The corresponding line element of the metric reads

```
ds3 = (1-m/2/rho)^2/(1 + m/2/rho)^2 Dt[t]^2 -
(1 + m/2/rho)^4 ( Dt[rho]^2 + rho^2 ( Dt[th]^2
+ Sin[th]^2 Dt[phi]^2))

        m    2       2
(1 - -----)   Dt[t]
     2 rho                         m    4
------------------- - (1 + -----)
         m    2                   2 rho
   (1 + -----)
         2 rho

          2      2       2       2        2
   (Dt[rho]  + rho  (Dt[th]  + Dt[phi]  Sin[th] ))
```

and the corresponding metric tensor is

```
g = metric[ds3,{t,rho,th,phi}]

          2
         m                      4 m rho
{{-------------------- - -------------------- +
   2                     2              2
  m  + 4 m rho + 4 rho   m  + 4 m rho + 4 rho

          2
      4 rho
-------------------, 0, 0, 0},
   2                 2
  m  + 4 m rho + 4 rho

          4      3        2
         m      m      3 m     2 m
{0, -1 - ------- - ------ - ------ - ---, 0, 0},
            4        3        2     rho
        16 rho   2 rho    2 rho

       2      4       3
   -3 m      m       m                        2
{0, 0, ----- - ------- - ----- - 2 m rho - rho , 0},
        2          2     2 rho
            16 rho
```

```
                  2           2   4          2   3        2
          -3 m  Sin[th]      m  Sin[th]    m  Sin[th]
{0, 0, 0, --------------  - -----------  - ----------- -
                  2              2            2 rho
              16 rho

             2       2        2    2      2
    2 m rho Sin[th]   - rho  Sin[th] }}
```

Up to now we have only discussed the line element of the metric and its related metric tensor. To derive the field equations for the vacuum case in GR, we have to introduce other tensors. One of the essential quantities determining the field equations are the Christoffel symbols. These symbols are related to the metric tensor in a straightforward way.

6.3.2 The Christoffel symbols

Every important relation or equation in a Riemannian space can be expressed in terms of the metric tensor or its partial derivatives. These expressions are often very complex. The Christoffel symbols are important expressions for formulating Einstein's field equations and for expressing the geometric properties of space. The Christoffel symbols contain the inverse of the metric tensor **ginv** and partial derivatives of first order with respect to the coordinates. The Christoffel symbols can be defined in *Mathematica* by

```
Christoffel[m_,a_,b_,g_,ginv_]:= Block[{n},
        Expand[
        Sum[ginv[[m,n]] (D[g[[a,n]],IndepVar[[b]] ] +
                         D[g[[b,n]],IndepVar[[a]] ] -
                         D[g[[a,b]],IndepVar[[n]] ]),
                    {n,1,Length[g]}]/2]
        ]
```

In mathematical notation the function **Christoffel[ ]** is given by

$$\Gamma_{a,b}^{m} = g^{mn}(\partial_b g_{an} + \partial_a g_{bn} - \partial_n g_{ab}). \tag{6.67}$$

Other important tensors which are needed to formulate the field equations are usually expressed in Christoffel symbols. The Christoffel symbols also appear in equations for metric geodesics (i.e., the equations defining the parameterized curve of a particle moving in space). In the following, we define tensors such as the Riemann tensor, the Ricci tensor, etc.

6.3.3 The Riemann tensor

The curvature tensor, also called the Riemann tensor, is defined in terms of Christoffel symbols by

```
Riemann[a_,b_,c_,d_,g_,ing_]:=Block[{},
        Expand[
        D[Christoffel[a,b,d,g,ing],IndepVar[[c]]] -
        D[Christoffel[a,b,c,g,ing],IndepVar[[d]]] +
        Sum[Christoffel[e,b,d,g,ing] Christoffel[a,e,c,g,ing],
            {e,1,Length[g]}] -
        Sum[Christoffel[e,b,c,g,ing] Christoffel[a,e,d,g,ing],
            {e,1,Length[g]}]
                                ]
        ]
```

The Riemann tensor describes the geometric properties of the underlying space. A flat space contains a Riemann tensor with zero coefficients.

A contraction of the Riemann tensor delivers the Ricci tensor. The Ricci tensor is a symmetric tensor in the form

```
Ricci[m_,q_,g_,ing_]:=Block[{a},
        Expand[
                Sum[Riemann[a,m,a,q,g,ing],
                {a,1,Length[g]}]]]
```

Another contraction of the Ricci tensor defines the curvature scalar or Ricci scalar

```
RicciScalar[g_,ing_]:= Block[{},
        Expand[Sum[ing[[a,b]]  Ricci[a,b,g,ing],
                {a,1,Length[g]},{b,1,Length[g]}]]];
```

Having these tensors available, we can proceed to the derivation of Einstein's field equations.

6.3.4 Einstein's field equations

Einstein's vacuum equations are expressed by the Ricci tensor and the Ricci scalar

```
Einstein[m_,n_,g_,ing_]:=Ricci[m,n,g,ing] -
                RicciScalar[g,ing] g[[m,n]]/2
```

The function **Einstein**[] gives the left hand side of the equations while the right hand side is equal to zero. The derived equations are nonlinear partial differential equations of second order in space and time. In addition to the field equations, there are four side conditions given by the Bianchi identities; these identities are a form of energy conservation.

```
Bianchi[a_,g_,ing_]:=Block[ {},
Expand[
Sum[ D[Sum[ ing[[n,m]] Einstein[m,a,g.ing],
   {m,1,Length[g]}],IndepVar[[n]] ],{n,1,Length[g]}]
+ Sum[ Sum[ Christoffel[n,m,n,g,ing]
  Sum[ ing[[m,l]] Einstein[l,a,g,ing],
   {l,1,Length[g]}],{m,1,Length[g]}],{n,1,Length[g]}]
- Sum[ Sum[ Christoffel[n,m,a,g,ing]
  Sum[ ing[[m,l]] Einstein[l,n,g,ing],
   {l,1,Length[g]}],{m,1,Length[g]}],{n,1,Length[g]}]  ]
];
```

The calculation of the ten coefficients of the metric tensor g is an incompletely formulated mathematical problem since there are fewer equations than unknowns (six equations with ten unknowns). Since the metric tensor is a solution of the field equations, it is apparent that a coordinate transformation does not change the problem. When choosing a coordinate system, we are free to introduce gauge conditions. For example, Gaussian or normal coordinates are often introduced by setting $g_{00} = 1$ and $g_{0a} = 0$ for $a = 1, 2, 3$.

We now examine some examples for which we can use the functions defined above. The first is again the three-dimensional flat Cartesian space.

6.3.5 The Cartesian space

The Cartesian space in three dimensions is characterized by the line element

```
dsc = Dt[x]^2 + Dt[y]^2 + Dt[z]^2

       2        2        2
Dt[x]   + Dt[y]   + Dt[z]
```

with the independent variables

```
IndepVar = {x,y,z}

{x, y, z}
```

The metric form of this space is given by

```
g = metric[dsc,IndepVar]

{{1, 0, 0}, {0, 1, 0}, {0, 0, 1}}
```

The inverse of the metric tensor follows by

```
ing = Inverse[g]

{{1, 0, 0}, {0, 1, 0}, {0, 0, 1}}
```

which is simply the identity matrix. Then, we calculate some of the Christoffel symbols to see which of them is not equal to zero.

```
Christoffel[1,1,1,g,ing]

0

Christoffel[1,2,1,g,ing]

0

Ricci[1,2,g,ing]

0
```

It is trivial to see that all Christoffel symbols of this metric vanish. Consequently, the coefficients of the Riemann tensor vanish, too. This fact is expected because a Cartesian space is flat. We now examine the Cartesian space in different coordinate systems.

6.3.6 Cartesian space in cylindrical coordinates

The line element of Cartesian space with cylindrical coordinates is expressed by

```
IndepVar = {r,phi,z}

{r, phi, z}

dscy = Dt[r]^2 + r^2 Dt[phi]^2 + Dt[z]^2
```

```
  2           2         2          2
 r  Dt[phi]   + Dt[r]   + Dt[z]
```

The metric tensor is given by

```
 g = metric[dscy,IndepVar]

                2
 {{1, 0, 0}, {0, r , 0}, {0, 0, 1}}
```

and the inverse of the metric tensor is

```
 ing = Inverse[g]

                -2
 {{1, 0, 0}, {0, r  , 0}, {0, 0, 1}}
```

Contrary to the case of the Cartesian coordinate system, the Christoffel symbols do not all vanish.

```
 Table[Christoffel[i,j,k,g,ing],{i,1,3},{j,1,3},
      {k,1,3}] // MatrixForm

 0    0    0


 0    -r   0


 0    0    0


       1
       -
 0     r    0
 1
 -
 r    0    0


 0    0    0


 0    0    0
```

```
0    0    0

0    0    0
```

Nevertheless, the Riemann tensor has to be zero for flat Cartesian space in spite of the coordinate transformation.

```
Table[Riemann[a,b,c,d,g,ing],{a,1,3},{b,1,3},
    {c,1,3},{d,1,3}]

{{{{0, 0, 0}, {0, 0, 0}, {0, 0, 0}},

   {{0, 0, 0}, {0, 0, 0}, {0, 0, 0}},

   {{0, 0, 0}, {0, 0, 0}, {0, 0, 0}}},

  {{{0, 0, 0}, {0, 0, 0}, {0, 0, 0}},

   {{0, 0, 0}, {0, 0, 0}, {0, 0, 0}},

   {{0, 0, 0}, {0, 0, 0}, {0, 0, 0}}},

  {{{0, 0, 0}, {0, 0, 0}, {0, 0, 0}},

   {{0, 0, 0}, {0, 0, 0}, {0, 0, 0}},

   {{0, 0, 0}, {0, 0, 0}, {0, 0, 0}}}}}
```

The disappearance of the Riemann tensor in flat Cartesian space is independent of the corresponding coordinate system. To illustrate the situation we examine next the Euclidean space in polar coordinates.

6.3.7 Euclidean space in polar coordinates

With the spherical coordinates

```
IndepVar = {r,theta,phi}

{r, theta, phi}
```

the line element and the corresponding metric are given by

```
dscp = Dt[r]^2 + r^2 Dt[theta]^2 +
                 r^2 Sin[theta]^2 Dt[phi]^2

      2    2          2    2        2          2
Dt[r]  + r  Dt[theta]  + r  Dt[phi]  Sin[theta]

g = metric[dscp,IndepVar]

                2          2          2          2
{{1, 0, 0}, {0, r , 0}, {0, 0, r  Sin[theta] }}
```

The inverse metric tensor is

```
ing = Inverse[g]
                                        2
                  -2            Csc[theta]
{{1, 0, 0}, {0, r  , 0}, {0, 0, ----------}}
                                     2
                                    r
```

The Christoffel symbols read

```
Table[Christoffel[i,j,k,g,ing],{i,1,3},{j,1,3},{k,1,3}]

                                          2
{{{0, 0, 0}, {0, -r, 0}, {0, 0, -(r Sin[theta] )}},

       1        1
  {{0, -, 0}, {-, 0, 0},
       r        r

    {0, 0, -(Cos[theta] Sin[theta])}},

       1              1
  {{0, 0, -}, {0, 0, Cot[theta]}, {-, Cot[theta], 0}}}
       r              r
```

As in the previous example, the Christoffel symbols do not vanish — and are now even more complicated. But again, as expected, the coefficients of the Riemann tensor are zero.

```
Simplify[Table[Riemann[a,b,c,d,g,ing],
{a,1,3},{b,1,3},{c,1,3},{d,1,3}] ] // MatrixForm

0 0 0   0 0 0   0 0 0
0 0 0   0 0 0   0 0 0
0 0 0   0 0 0   0 0 0

0 0 0   0 0 0   0 0 0
0 0 0   0 0 0   0 0 0
0 0 0   0 0 0   0 0 0

0 0 0   0 0 0   0 0 0
0 0 0   0 0 0   0 0 0
0 0 0   0 0 0   0 0 0
```

6.4 The Schwarzschild solution

In this section, we discuss a nontrivial solution of Einstein's field equations, the famous Schwarzschild metric given in special coordinate representations. The Schwarzschild solution is a solution of Einstein's field equations with the highest symmetry, i.e., with spherical symmetry.

6.4.1 The Schwarzschild metric in Eddington-Finkelstein form

In this representation, there are as usual a time-like variable t, a variable r related to distance, and two angle variables θ and ϕ.

```
IndepVar = {t,r,theta,phi}

{t, r, theta, phi}
```

According to the Eddington-Finkelstein line element

```
dss = (1-2m/r) Dt[t]^2 - 4m/r Dt[t] Dt[r] -
      (1+2m/r) Dt[r]^2 - r^2(Dt[theta]^2 +
      Sin[theta]^2 Dt[phi]^2)

        2 m        2    4 m Dt[r] Dt[t]        2 m         2
-((1 + ---) Dt[r] ) - --------------- + (1 - ---) Dt[t]  -
        r                    r                 r

    2            2            2            2
   r  (Dt[theta]  + Dt[phi]  Sin[theta] )
```

The meaning of r and t is different from the standard Schwarzschild solution. Due to our choice of r, a nondiagonal element between r and t appears. Here, the diagonal elements of r and t are in a more symmetric form. Yet, the metric possesses the required symmetries, spherical symmetry and time independence. This metric is special in that it is regular at point $r = 2m$ whereas the Schwarzschild line element in its standard form is singular at this point. This metric can be interpreted as an analytical extension of the standard form in the region $2m < r < \infty$ to the region $0 < r < \infty$. With the metric tensor

```
g = metric[dss,IndepVar]

        2 m  -2 m          -2 m       2 m
{{1 - ---, ----, 0, 0}, {----, -1 - ---, 0, 0},
       r    r             r          r

            2              2             2
   {0, 0, -r , 0}, {0, 0, 0, -(r  Sin[theta] )}}
```

and its inverse

```
ing = Inverse[g]

           2          3       2    4           2
   Csc[theta]  (-2 m r  Sin[theta]  - r  Sin[theta] )
{{-(---------------------------------------------------),
                          4
                         r

  -2 m          -2 m
  ----, 0, 0}, {----, -(
   r             r

           2          3       2    4           2
   Csc[theta]  (-2 m r  Sin[theta]  + r  Sin[theta] )
   --------------------------------------------------),
                         4
                        r

                                              2
              -2                    Csc[theta]
  0, 0}, {0, 0, -r  , 0}, {0, 0, 0, -(-----------)}}
                                          2
                                         r
```

the Christoffel symbols and Ricci tensor are easily calculated.

```
Table[Christoffel[i,j,k,g,ing],{i,1,4},{j,1,4},{k,1,4}]

        2    2                      2        2
    2 m    2 m    m              2 m    m    2 m    2 m
{{{----, ---- + --, 0, 0}, {---- + --, ---- + ---, 0, 0},
     3    3     2              3     2    3      2
    r    r     r              r     r    r      r


                                                    2
  {0, 0, -2 m, 0}, {0, 0, 0, -2 m Sin[theta] }},


        2             2                  2       2
   -2 m    m     -2 m              -2 m    -2 m    m
{{----- + --, -----, 0, 0}, {-----, ----- - --, 0, 0},
     3    2     3              3       3     2
    r    r     r              r       r     r


  {0, 0, 2 m - r, 0}, {0, 0, 0,

              2             2
    2 m Sin[theta]  - r Sin[theta] }},

                      1         1
  {{0, 0, 0, 0}, {0, 0, -, 0}, {0, -, 0, 0},
                      r         r

  {0, 0, 0, -(Cos[theta] Sin[theta])}},

                         1
  {{0, 0, 0, 0}, {0, 0, 0, -}, {0, 0, 0, Cot[theta]},
                         r

      1
  {0, -, Cot[theta], 0}}}
      r
```

```
Table[Ricci[i,j,g,ing],{i,1,4},{j,1,4}]

{{0, 0, 0, 0}, {0, 0, 0, 0},

                    2             2
  {0, 0, -1 - Cot[theta]  + Csc[theta] , 0}, {0, 0, 0, 0}}
```

With these quantities in hand we can verify that the form of the Eddington–Finkelstein line element is a solution of Einstein's vacuum field equations:

```
Simplify[ Table[Einstein[a,b,g,ing],{a,1,4},{b,1,4}] ]

{{0, 0, 0, 0}, {0, 0, 0, 0}, {0, 0, 0, 0}, {0, 0, 0, 0}}
```

In addition to the field equations the Bianchi identities are satisfied, too.

6.4.2 Dingle's metric

The metric of Dingle with three space coordinates and one time-like coordinate

```
IndepVar = {t,x,y,z}

{t, x, y, z}
```

is the most general metric in diagonal form.

```
dsd = A1[t,x,y,z] Dt[t]^2 -
      B1[t,x,y,z] Dt[x]^2 -
      C1[t,x,y,z] Dt[y]^2 -
      D1[t,x,y,z] Dt[z]^2

                     2                          2
 A1[t, x, y, z] Dt[t]   - B1[t, x, y, z] Dt[x]   -

                      2        2
   C1[t, x, y, z] Dt[y]  - Dt[z]  D1[t, x, y, z]
```

Hence the metric tensor is a diagonal tensor

```
g = metric[dsd,IndepVar]

{{A1[t, x, y, z], 0, 0, 0}, {0, -B1[t, x, y, z], 0, 0},

  {0, 0, -C1[t, x, y, z], 0}, {0, 0, 0, -D1[t, x, y, z]}}
```

and so is its inverse

```
ing = Inverse[g]

          1                               1
{{-------------, 0, 0, 0}, {0, -(-------------), 0, 0},
   A1[t, x, y, z]                   B1[t, x, y, z]


              1
 {0, 0, -(-------------), 0},
           C1[t, x, y, z]


                 1
 {0, 0, 0, -(-------------)}}
             D1[t, x, y, z]
```

Due to the form of the metric tensor, the Christoffel symbols are fairly simple expressions.

```
Table[christoffel[i,j,k,g,ing],{i,1,4},{j,1,4},{k,1,4}]


         (1,0,0,0)                    (0,1,0,0)
      A1          [t, x, y, z]   A1          [t, x, y, z]
{{{----------------------,      ----------------------,
      2 A1[t, x, y, z]            2 A1[t, x, y, z]


         (0,0,1,0)                    (0,0,0,1)
      A1          [t, x, y, z]   A1          [t, x, y, z]
   ----------------------,      ----------------------},
      2 A1[t, x, y, z]            2 A1[t, x, y, z]


         (0,1,0,0)                    (1,0,0,0)
      A1          [t, x, y, z]   B1          [t, x, y, z]
 {----------------------,      ----------------------, 0,
      2 A1[t, x, y, z]            2 A1[t, x, y, z]


             (0,0,1,0)
          A1          [t, x, y, z]
  0}, {----------------------, 0,
          2 A1[t, x, y, z]


         (1,0,0,0)
      C1          [t, x, y, z]
   ----------------------, 0},
      2 A1[t, x, y, z]


         (0,0,0,1)                        (1,0,0,0)
      A1          [t, x, y, z]         D1          [t, x, y, z]
 {----------------------, 0, 0,      ----------------------}}
      2 A1[t, x, y, z]                 2 A1[t, x, y, z]
```

```
          (0,1,0,0)                (1,0,0,0)
        A1            [t, x, y, z]  B1           [t, x, y, z]
 , {{---------------------------,  ----------------------------, 0,
        2 B1[t, x, y, z]              2 B1[t, x, y, z]

          (1,0,0,0)                  (0,1,0,0)
        B1            [t, x, y, z]  B1            [t, x, y, z]
 0}, {---------------------------,  ----------------------------,
        2 B1[t, x, y, z]              2 B1[t, x, y, z]

     (0,0,1,0)                   (0,0,0,1)
   B1            [t, x, y, z]  B1            [t, x, y, z]
  ----------------------------, ----------------------------},
     2 B1[t, x, y, z]             2 B1[t, x, y, z]

          (0,0,1,0)                  (0,1,0,0)
        B1            [t, x, y, z]  -C1           [t, x, y, z]
  {0, ---------------------------,  ----------------------------,
        2 B1[t, x, y, z]              2 B1[t, x, y, z]

             (0,0,0,1)
           B1            [t, x, y, z]
  0}, {0, ----------------------------, 0,
           2 B1[t, x, y, z]

     (0,1,0,0)
   -D1           [t, x, y, z]
  ----------------------------}},
     2 B1[t, x, y, z]

          (0,0,1,0)                  (1,0,0,0)
        A1            [t, x, y, z]  C1           [t, x, y, z]
{{---------------------------, 0,  ----------------------------,
        2 C1[t, x, y, z]              2 C1[t, x, y, z]

             (0,0,1,0)
           -B1           [t, x, y, z]
  0}, {0, ----------------------------,
           2 C1[t, x, y, z]

     (0,1,0,0)
   C1            [t, x, y, z]
  ----------------------------, 0},
     2 C1[t, x, y, z]

          (1,0,0,0)                  (0,1,0,0)
        C1            [t, x, y, z]  C1            [t, x, y, z]
 {----------------------------,  ----------------------------,
     2 C1[t, x, y, z]               2 C1[t, x, y, z]

     (0,0,1,0)                   (0,0,0,1)
   C1            [t, x, y, z]  C1            [t, x, y, z]
  ----------------------------, ----------------------------},
     2 C1[t, x, y, z]             2 C1[t, x, y, z]
```

```
            (0,0,0,1)
        C1              [t, x, y, z]
  {0, 0, -----------------------,
            2 C1[t, x, y, z]



     (0,0,1,0)
   -D1              [t, x, y, z]
  -----------------------}},
     2 C1[t, x, y, z]



     (0,0,0,1)
   A1              [t, x, y, z]
  {{-----------------------, 0, 0,
     2 D1[t, x, y, z]



     (1,0,0,0)
   D1              [t, x, y, z]
  -----------------------},
     2 D1[t, x, y, z]



            (0,0,0,1)
      -B1              [t, x, y, z]
  {0, -----------------------, 0,
            2 D1[t, x, y, z]



     (0,1,0,0)
   D1              [t, x, y, z]
  -----------------------},
     2 D1[t, x, y, z]



            (0,0,0,1)
        -C1              [t, x, y, z]
  {0, 0, -----------------------,
            2 D1[t, x, y, z]



     (0,0,1,0)
   D1              [t, x, y, z]
  -----------------------},
     2 D1[t, x, y, z]


     (1,0,0,0)                    (0,1,0,0)
   D1              [t, x, y, z]  D1              [t, x, y, z]
  {-----------------------,  -----------------------,
     2 D1[t, x, y, z]            2 D1[t, x, y, z]


     (0,0,1,0)                    (0,0,0,1)
   D1              [t, x, y, z]  D1              [t, x, y, z]
  -----------------------,  -----------------------}}}
     2 D1[t, x, y, z]            2 D1[t, x, y, z]
```

Still, even one equation of Einstein's vacuum field equations is complicated

```
Einstein[1,1,g,ing]

      (0,0,0,1)        2
    -A1        [t, x, y, z]
  ----------------------------- +
  4 A1[t, x, y, z] D1[t, x, y, z]

    (0,0,0,1)               (0,0,0,1)
  A1        [t, x, y, z] B1        [t, x, y, z]
  ------------------------------------------------ +
        4 B1[t, x, y, z] D1[t, x, y, z]

    (0,0,0,1)               (0,0,0,1)
  A1        [t, x, y, z] C1        [t, x, y, z]
  ------------------------------------------------ -
        4 C1[t, x, y, z] D1[t, x, y, z]

    (0,0,0,1)               (0,0,0,1)
  A1        [t, x, y, z] D1        [t, x, y, z]
  ------------------------------------------------ +
                          2
              4 D1[t, x, y, z]

    (0,0,0,2)
  A1        [t, x, y, z]
  ---------------------- -
      2 D1[t, x, y, z]

      (0,0,1,0)        2
    A1        [t, x, y, z]
  ----------------------------- +
  4 A1[t, x, y, z] C1[t, x, y, z]

    (0,0,1,0)               (0,0,1,0)
  A1        [t, x, y, z] B1        [t, x, y, z]
  ------------------------------------------------ -
        4 B1[t, x, y, z] C1[t, x, y, z]

    (0,0,1,0)               (0,0,1,0)
  A1        [t, x, y, z] C1        [t, x, y, z]
  ------------------------------------------------ +
                          2
              4 C1[t, x, y, z]

    (0,0,1,0)               (0,0,1,0)
  A1        [t, x, y, z] D1        [t, x, y, z]
  ------------------------------------------------ +
        4 C1[t, x, y, z] D1[t, x, y, z]

    (0,0,2,0)
  A1        [t, x, y, z]
  ---------------------- -
      2 C1[t, x, y, z]
```

```
     (0,1,0,0)              2
  A1         [t, x, y, z]
------------------------------ -
4 A1[t, x, y, z] B1[t, x, y, z]


  (0,1,0,0)                (0,1,0,0)
A1         [t, x, y, z] B1         [t, x, y, z]
--------------------------------------------- +
                                    2
                 4 B1[t, x, y, z]


  (0,1,0,0)                (0,1,0,0)
A1         [t, x, y, z] C1         [t, x, y, z]
--------------------------------------------- +
          4 B1[t, x, y, z] C1[t, x, y, z]


  (0,1,0,0)                (0,1,0,0)
A1         [t, x, y, z] D1         [t, x, y, z]
--------------------------------------------- +
          4 B1[t, x, y, z] D1[t, x, y, z]

  (0,2,0,0)
A1         [t, x, y, z]
---------------------- +
    2 B1[t, x, y, z]

  (1,0,0,0)                (1,0,0,0)
A1         [t, x, y, z] B1         [t, x, y, z]
--------------------------------------------- +
          4 A1[t, x, y, z] B1[t, x, y, z]


  (1,0,0,0)              2
B1         [t, x, y, z]
----------------------- +
                 2
    4 B1[t, x, y, z]

  (1,0,0,0)                (1,0,0,0)
A1         [t, x, y, z] C1         [t, x, y, z]
--------------------------------------------- +
          4 A1[t, x, y, z] C1[t, x, y, z]

  (1,0,0,0)              2
C1         [t, x, y, z]
----------------------- +
                 2
    4 C1[t, x, y, z]

  (1,0,0,0)                (1,0,0,0)
A1         [t, x, y, z] D1         [t, x, y, z]
--------------------------------------------- +
          4 A1[t, x, y, z] D1[t, x, y, z]
```

$$\frac{D1^{(1,0,0,0)}[t,\ x,\ y,\ z]^2}{4\ D1[t,\ x,\ y,\ z]^2} - \frac{B1^{(2,0,0,0)}[t,\ x,\ y,\ z]}{2\ B1[t,\ x,\ y,\ z]} -$$

$$\frac{C1^{(2,0,0,0)}[t,\ x,\ y,\ z]}{2\ C1[t,\ x,\ y,\ z]} - \frac{D1^{(2,0,0,0)}[t,\ x,\ y,\ z]}{2\ D1[t,\ x,\ y,\ z]} -$$

$$A1[t,\ x,\ y,\ z]\ (\frac{-A1^{(0,0,0,1)}[t,\ x,\ y,\ z]^2}{2\ A1[t,\ x,\ y,\ z]^2\ D1[t,\ x,\ y,\ z]} +$$

$$\frac{A1^{(0,0,0,1)}[t,\ x,\ y,\ z]\ B1^{(0,0,0,1)}[t,\ x,\ y,\ z]}{2\ A1[t,\ x,\ y,\ z]\ B1[t,\ x,\ y,\ z]\ D1[t,\ x,\ y,\ z]} -$$

$$\frac{B1^{(0,0,0,1)}[t,\ x,\ y,\ z]^2}{2\ B1[t,\ x,\ y,\ z]^2\ D1[t,\ x,\ y,\ z]} +$$

$$\frac{A1^{(0,0,0,1)}[t,\ x,\ y,\ z]\ C1^{(0,0,0,1)}[t,\ x,\ y,\ z]}{2\ A1[t,\ x,\ y,\ z]\ C1[t,\ x,\ y,\ z]\ D1[t,\ x,\ y,\ z]} +$$

$$\frac{B1^{(0,0,0,1)}[t,\ x,\ y,\ z]\ C1^{(0,0,0,1)}[t,\ x,\ y,\ z]}{2\ B1[t,\ x,\ y,\ z]\ C1[t,\ x,\ y,\ z]\ D1[t,\ x,\ y,\ z]} -$$

$$\frac{C1^{(0,0,0,1)}[t,\ x,\ y,\ z]^2}{2\ C1[t,\ x,\ y,\ z]^2\ D1[t,\ x,\ y,\ z]} -$$

$$\frac{A1^{(0,0,0,1)}[t,\ x,\ y,\ z]\ D1^{(0,0,0,1)}[t,\ x,\ y,\ z]}{2\ A1[t,\ x,\ y,\ z]\ D1[t,\ x,\ y,\ z]^2} -$$

$$\frac{B1^{(0,0,0,1)}[t,\ x,\ y,\ z]\ D1^{(0,0,0,1)}[t,\ x,\ y,\ z]}{2\ B1[t,\ x,\ y,\ z]\ D1[t,\ x,\ y,\ z]^2} -$$

```
    (0,0,0,1)                    (0,0,0,1)
C1          [t, x, y, z] D1          [t, x, y, z]
--------------------------------------------- +
                                      2
        2 C1[t, x, y, z] D1[t, x, y, z]


      (0,0,0,2)
  A1          [t, x, y, z]
---------------------------- +
A1[t, x, y, z] D1[t, x, y, z]


      (0,0,0,2)
  B1          [t, x, y, z]
---------------------------- +
B1[t, x, y, z] D1[t, x, y, z]


      (0,0,0,2)
  C1          [t, x, y, z]
---------------------------- -
C1[t, x, y, z] D1[t, x, y, z]


       (0,0,1,0)            2
   A1          [t, x, y, z]
-------------------------------- +
                2
2 A1[t, x, y, z]  C1[t, x, y, z]


  (0,0,1,0)                    (0,0,1,0)
A1          [t, x, y, z] B1          [t, x, y, z]
--------------------------------------------- -
2 A1[t, x, y, z] B1[t, x, y, z] C1[t, x, y, z]


       (0,0,1,0)            2
   B1          [t, x, y, z]
-------------------------------- -
                2
2 B1[t, x, y, z]  C1[t, x, y, z]


  (0,0,1,0)                  (0,0,1,0)
A1          [t, x, y, z] C1          [t, x, y, z]
--------------------------------------------- -
                                      2
        2 A1[t, x, y, z] C1[t, x, y, z]


  (0,0,1,0)                    (0,0,1,0)
B1          [t, x, y, z] C1          [t, x, y, z]
--------------------------------------------- +
                                      2
        2 B1[t, x, y, z] C1[t, x, y, z]


  (0,0,1,0)                    (0,0,1,0)
A1          [t, x, y, z] D1          [t, x, y, z]
--------------------------------------------- +
2 A1[t, x, y, z] C1[t, x, y, z] D1[t, x, y, z]
```

```
   (0,0,1,0)                  (0,0,1,0)
B1        [t, x, y, z] D1         [t, x, y, z]
--------------------------------------------- -
2 B1[t, x, y, z] C1[t, x, y, z] D1[t, x, y, z]

   (0,0,1,0)                  (0,0,1,0)
C1        [t, x, y, z] D1         [t, x, y, z]
--------------------------------------------- -
                   2
        2 C1[t, x, y, z]  D1[t, x, y, z]

     (0,0,1,0)            2
   D1        [t, x, y, z]
------------------------------- +
                            2
2 C1[t, x, y, z] D1[t, x, y, z]

     (0,0,2,0)
   A1        [t, x, y, z]
--------------------------- +
A1[t, x, y, z] C1[t, x, y, z]

     (0,0,2,0)
   B1        [t, x, y, z]
--------------------------- +
B1[t, x, y, z] C1[t, x, y, z]

     (0,0,2,0)
   D1        [t, x, y, z]
--------------------------- -
C1[t, x, y, z] D1[t, x, y, z]

     (0,1,0,0)            2
   A1        [t, x, y, z]
------------------------------- -
                   2
2 A1[t, x, y, z]  B1[t, x, y, z]

  (0,1,0,0)                (0,1,0,0)
A1        [t, x, y, z] B1         [t, x, y, z]
--------------------------------------------- +
                             2
        2 A1[t, x, y, z] B1[t, x, y, z]

  (0,1,0,0)                  (0,1,0,0)
A1        [t, x, y, z] C1          [t, x, y, z]
--------------------------------------------- -
2 A1[t, x, y, z] B1[t, x, y, z] C1[t, x, y, z]


  (0,1,0,0)                  (0,1,0,0)
B1        [t, x, y, z] C1          [t, x, y, z]
--------------------------------------------- -
                   2
        2 B1[t, x, y, z]  C1[t, x, y, z]
```

```
        (0,1,0,0)              2
    C1           [t, x, y, z]
-------------------------------- +
                          2
2 B1[t, x, y, z] C1[t, x, y, z]


  (0,1,0,0)                  (0,1,0,0)
A1         [t, x, y, z] D1           [t, x, y, z]
------------------------------------------------ -
2 A1[t, x, y, z] B1[t, x, y, z] D1[t, x, y, z]


  (0,1,0,0)                  (0,1,0,0)
B1         [t, x, y, z] D1           [t, x, y, z]
------------------------------------------------ +
                          2
      2 B1[t, x, y, z]  D1[t, x, y, z]


  (0,1,0,0)                  (0,1,0,0)
C1         [t, x, y, z] D1           [t, x, y, z]
------------------------------------------------ -
2 B1[t, x, y, z] C1[t, x, y, z] D1[t, x, y, z]


      (0,1,0,0)              2
    D1           [t, x, y, z]
-------------------------------- +
                          2
2 B1[t, x, y, z] D1[t, x, y, z]


      (0,2,0,0)
    A1           [t, x, y, z]
---------------------------- +
A1[t, x, y, z] B1[t, x, y, z]


      (0,2,0,0)
    C1           [t, x, y, z]
---------------------------- +
B1[t, x, y, z] C1[t, x, y, z]


      (0,2,0,0)
    D1           [t, x, y, z]
---------------------------- +
B1[t, x, y, z] D1[t, x, y, z]


  (1,0,0,0)                  (1,0,0,0)
A1         [t, x, y, z] B1           [t, x, y, z]
------------------------------------------------ +
                          2
      2 A1[t, x, y, z]  B1[t, x, y, z]


      (1,0,0,0)              2
    B1           [t, x, y, z]
-------------------------------- +
                          2
2 A1[t, x, y, z] B1[t, x, y, z]
```

```
     (1,0,0,0)                 (1,0,0,0)
 A1           [t, x, y, z] C1           [t, x, y, z]
-------------------------------------------------- -
                            2
         2 A1[t, x, y, z]  C1[t, x, y, z]

     (1,0,0,0)                 (1,0,0,0)
 B1           [t, x, y, z] C1           [t, x, y, z]
-------------------------------------------------- +
       2 A1[t, x, y, z] B1[t, x, y, z] C1[t, x, y, z]

       (1,0,0,0)               2
     C1           [t, x, y, z]
------------------------------- +
                                2
     2 A1[t, x, y, z] C1[t, x, y, z]

     (1,0,0,0)                 (1,0,0,0)
 A1           [t, x, y, z] D1           [t, x, y, z]
-------------------------------------------------- -
                            2
         2 A1[t, x, y, z]  D1[t, x, y, z]

     (1,0,0,0)               (1,0,0,0)
 B1           [t, x, y, z] D1           [t, x, y, z]
-------------------------------------------------- -
       2 A1[t, x, y, z] B1[t, x, y, z] D1[t, x, y, z]

     (1,0,0,0)               (1,0,0,0)
 C1           [t, x, y, z] D1           [t, x, y, z]
-------------------------------------------------- +
       2 A1[t, x, y, z] C1[t, x, y, z] D1[t, x, y, z]


       (1,0,0,0)              2
     D1           [t, x, y, z]
------------------------------- -
                                2
     2 A1[t, x, y, z] D1[t, x, y, z]

       (2,0,0,0)
     B1           [t, x, y, z]
----------------------------- -
     A1[t, x, y, z] B1[t, x, y, z]

       (2,0,0,0)
     C1           [t, x, y, z]
----------------------------- -
     A1[t, x, y, z] C1[t, x, y, z]

       (2,0,0,0)
     D1           [t, x, y, z]
-----------------------------)) / 2
     A1[t, x, y, z] D1[t, x, y, z]
```

6.4.3 Schwarzschild metric in Kruskal coordinates

The Kruskal solution is the most general analytical extension of the Schwarzschild metric. While the Eddington–Finkelstein solution is developed for the time region $0 \leq t < \infty$ or $-\infty < t \leq 0$, the Kruskal solution is extended to both time regions.

The Kruskal solution consists of the two angle variables *theta* and *phi*, a space-like variable x and a time-like variable t.

```
IndepVar = {t,x,theta,phi}

{t, x, theta, phi}
```

The radial distance r is defined implicitly by the equation.

```
gld = t^2 - x^2 == - (r[x,t] - 2 m) Exp[r[x,t]/(2 m)]

  2    2         r[x, t]/(2 m)
 t  - x  == -(E              (-2 m + r[x, t]))
```

For later calculations this equation is solved for **t**.

```
seq = Solve[ gld /. r[x,t] -> r ,t] //Last

                r/m              r/m      2
{t -> Sqrt[2 Sqrt[E   ] m - Sqrt[E   ] r + x ]}
```

The line element is given by the radial coordinate r

```
dsk = 16 m^2/r[x,t] Exp[-r[x,t]/(2m)] Dt[t]^2 -
      16 m^2/r[x,t] Exp[-r[x,t]/(2m)] Dt[x]^2 -
      r[x,t]^2 ( Dt[theta]^2 + Sin[theta]^2 Dt[phi]^2)

         2       2                    2       2
   16 m   Dt[t]                 16 m   Dt[x]
 --------------------- - --------------------- -
  r[x, t]/(2 m)           r[x, t]/(2 m)
 E            r[x, t]    E            r[x, t]

          2         2        2          2
   r[x, t]  (Dt[theta]  + Dt[phi]  Sin[theta] )
```

The metric is again in the shape of a diagonal matrix and its inverse

```
g = metric[dsk,IndepVar]
```

$$
\left\{\left\{\frac{16 \; \mathrm{Sqrt}[E^{-(r[x,\;t]/m)}] \; m^2}{r[x,\;t]}, \; 0, \; 0, \; 0\right\},\right.
$$

$$
\left\{0, \; \frac{-16 \; \mathrm{Sqrt}[E^{-(r[x,\;t]/m)}] \; m^2}{r[x,\;t]}, \; 0, \; 0\right\},
$$

$$
\left.\{0, \; 0, \; -r[x,\;t]^2, \; 0\}, \; \{0, \; 0, \; 0, \; -(r[x,\;t]^2 \; \mathrm{Sin}[theta]^2)\}\right\}\right\}
$$

```
ing = Inverse[g]
```

$$
\left\{\left\{\frac{E^{r[x,\;t]/m} \; \mathrm{Sqrt}[E^{-(r[x,\;t]/m)}] \; r[x,\;t]}{16 \; m^2}, \; 0, \; 0, \; 0\right\},\right.
$$

$$
\left\{0, \; \frac{-(E^{r[x,\;t]/m} \; \mathrm{Sqrt}[E^{-(r[x,\;t]/m)}] \; r[x,\;t])}{16 \; m^2}, \; 0, \; 0\right\},
$$

$$
\left.\{0, \; 0, \; -r[x,\;t]^{-2}, \; 0\}, \; \left\{0, \; 0, \; 0, \; -\left(\frac{\mathrm{Csc}[theta]^2}{r[x,\;t]^2}\right)\right\}\right\}\right\}
$$

To calculate the Christoffel symbols and the Einstein tensor we compute the derivatives of **r[x,t]** up to second order following from equation **gld**.

```
s1= Flatten[Simplify[Solve[D[gld,x],D[r[x,t],x]]]];

s2 = Flatten[Simplify[Solve[D[gld,t],D[r[x,t],t]]]];

s3 = Flatten[Simplify[Solve[D[gld,x,x],D[r[x,t],x,x]] /.
        s1 ]];
```

```
s4 = Flatten[Simplify[Solve[D[gld,t,t],D[r[x,t],t,t]] /.
  s2 ]];

sg = Flatten[{s1,s2,s3,s4}]

    (1,0)                    4 m x
{r       [x, t] -> ----------------------,
                     r[x, t]/(2 m)
                   E                    r[x, t]


    (0,1)                    -4 m t
  r      [x, t] -> ----------------------,
                     r[x, t]/(2 m)
                   E                    r[x, t]


    (2,0)
  r      [x, t] ->

               2     2                r[x, t]/(2 m)         2
    4 m (-4 m x  - 2 x  r[x, t] + E              r[x, t] )
    ----------------------------------------------------,
                     r[x, t]/m          3
                   E              r[x, t]


    (0,2)
  r      [x, t] ->

                2     2               r[x, t]/(2 m)         2
    -4 m (4 m t  + 2 t  r[x, t] + E              r[x, t] )
    ----------------------------------------------------}
                     r[x, t]/m          3
                   E              r[x, t]
```

With the list of **sg** rules the Christoffel symbols and the Einstein tensor are calculated as follows.

```
Table[ Simplify[ Christoffel[i,j,k,g,ing] /. sg ],
{i,1,4},{j,1,4},{k,1,4}]

        t (2 m + r[x, t])            x (2 m + r[x, t])
{{{----------------------, -(----------------------), 0,
     r[x, t]/(2 m)         2    r[x, t]/(2 m)         2
   E              r[x, t]     E              r[x, t]


               x (2 m + r[x, t])
    0}, {-(----------------------),
              r[x, t]/(2 m)         2
            E              r[x, t]


       t (2 m + r[x, t])
    ----------------------, 0, 0},
      r[x, t]/(2 m)         2
    E              r[x, t]
```

```
         r[x, t]/(2 m)          -(r[x, t]/m)
      -(E                Sqrt[E            ] t r[x, t])
 {0, 0, ---------------------------------------------,
                            4 m

                     r[x, t]/(2 m)        -(r[x, t]/m)
   0}, {0, 0, 0, -(E                Sqrt[E            ] t

                2
     r[x, t] Sin[theta] ) / (4 m)}},

          x (2 m + r[x, t])          t (2 m + r[x, t])
 {{-(----------------------), ----------------------, 0,
     r[x, t]/(2 m)      2     r[x, t]/(2 m)      2
    E            r[x, t]     E            r[x, t]

          t (2 m + r[x, t])
   0}, {----------------------,
        r[x, t]/(2 m)      2
       E            r[x, t]

          x (2 m + r[x, t])
   -(----------------------), 0, 0},
     r[x, t]/(2 m)      2
    E            r[x, t]

         r[x, t]/(2 m)          -(r[x, t]/m)
      -(E                Sqrt[E            ] x r[x, t])
 {0, 0, ---------------------------------------------,
                            4 m

                     r[x, t]/(2 m)        -(r[x, t]/m)
   0}, {0, 0, 0, -(E                Sqrt[E            ] x

                2
     r[x, t] Sin[theta] ) / (4 m)}},

                    -4 m t
 {{0, 0, ----------------------, 0},
         r[x, t]/(2 m)      2
        E            r[x, t]

                    4 m x
 {0, 0, ----------------------, 0},
        r[x, t]/(2 m)      2
       E            r[x, t]

             -4 m t                    4 m x
 {----------------------, ----------------------, 0,
  r[x, t]/(2 m)      2   r[x, t]/(2 m)      2
 E            r[x, t]   E            r[x, t]

                    -Sin[2 theta]
   0}, {0, 0, 0, -------------}},
                       2
```

```
                      -4 m t
{{0, 0, 0, ----------------------},
            r[x, t]/(2 m)        2
          E                r[x, t]

                       4 m x
 {0, 0, 0, ----------------------},
            r[x, t]/(2 m)        2
          E                r[x, t]

 {0, 0, 0, Cot[theta]},

         -4 m t                    4 m x
{----------------------, ----------------------,
  r[x, t]/(2 m)        2  r[x, t]/(2 m)        2
 E                r[x, t]  E              r[x, t]

 Cot[theta], 0}}}
```

To verify Einstein's field equations we calculate, for example, the (1,1) coefficient
of the Einstein tensor.

```
es1 = Simplify[ Einstein[1,1,g,ing] /. sg  ]

         r[x, t]/m        -(r[x, t]/m)       2   2
(8 m (2 E          Sqrt[E            ] m - t  + x  -

        r[x, t]/(2 m)             r[x, t]/m          3
      E              r[x, t])) / (E          r[x, t] )
```

With the aid of the defining equation for r, the above expression vanishes.

```
es1 = es1 /. { r[x,t] -> r}

        r/m        -(r/m)       r/(2 m)     2   2
8 m (2 E    Sqrt[E     ] m - E        r - t  + x )
--------------------------------------------------
                      r/m 3
                     E   r


Simplify[ PowerExpand[es1 /. seq ] ]

0
```

6.5 The Reissner–Nordstrom solution for a charged mass point

The Reissner–Nordstrom solution is a spherically symmetric metric for a massive body with charge ϵ. This type of solution allows the study of the coupling of Einstein's field equations with Maxwell's equations via the energy momentum tensor. Consequently, we have to solve the inhomogeneous field equations. Because of the spherical symmetry, we can use the Kruskal variables.

```
IndepVar = {t,r,theta,phi}

{t, r, theta, phi}
```

The same shape of the line element is also given.

```
dsr = Exp[nu[r]] Dt[t]^2 -
      Exp[lambda[r]] Dt[r]^2 -
      r^2 (Dt[theta]^2 + Sin[theta]^2 Dt[phi]^2)

     lambda[r]        2    nu[r]        2
 -(E          Dt[r] ) + E        Dt[t]  -

    2          2          2          2
   r  (Dt[theta]  + Dt[phi]  Sin[theta] )
```

```
g = metric[dsr,IndepVar]

      nu[r]                  lambda[r]              2
{{E        , 0, 0, 0}, {0, -E          , 0, 0}, {0, 0, -r , 0},

               2          2
    {0, 0, 0, -(r  Sin[theta] )}}
```

```
ing = Inverse[g]

    -nu[r]                    -lambda[r]
{{E        , 0, 0, 0}, {0, -E            , 0, 0},
```

```
                                        2
            -2                   Csc[theta]
  {0, 0, -r   , 0}, {0, 0, 0, -(-----------)}}
                                      2
                                     r
```

Since the Reissner–Nordstrom solution possesses spherical symmetry, the coordinates can be chosen so that the metric is static and *nu* and *lambda* depend only on the radial distance r. At the same time, the Reissner–Nordstrom solution satisfies Einstein's field equations and Maxwell's vacuum equations. Consequently the Maxwell tensor F also depends on the distance r. Its form is determined by a purely radial electrostatic field.

```
  F = {{ 0, - Ee[r],0,0},{Ee[r],0,0,0},{0,0,0,0},{0,0,0,0}}

  {{0, -Ee[r], 0, 0}, {Ee[r], 0, 0, 0}, {0, 0, 0, 0},

    {0, 0, 0, 0}}
```

According to Maxwell's equations, the covariant divergence of the Maxwell tensor must vanish. The conditions deliver the substitution rule

```
  sm = {Ee[r] -> epsilon Exp[ 1/2 ( nu[r] + lambda[r])]/
        r^2}

              (lambda[r] + nu[r])/2
          E                          epsilon
  {Ee[r] -> ------------------------------}
                         2
                        r
```

and the Maxwell tensor

```
  F = F /. sm

              (lambda[r] + nu[r])/2
          E                          epsilon
  {{0, -(------------------------------), 0, 0},
                       2
                      r
```

```
      (lambda[r] + nu[r])/2
   E                     epsilon
 {-----------------------------, 0, 0, 0}, {0, 0, 0, 0},
                2
               r

 {0, 0, 0, 0}}
```

with the corresponding covariant tensor.

```
 Fc = Simplify[ing . F . ing ]

      -lambda[r]/2 - nu[r]/2
   E                         epsilon
 {{0, -----------------------------, 0, 0},
                2
               r

      -lambda[r]/2 - nu[r]/2
    E                         epsilon
 {-(-----------------------------), 0, 0, 0},
                2
               r

 {0, 0, 0, 0}, {0, 0, 0, 0}}
```

The energy momentum tensor T is computed by

```
 T = Simplify[Table[ Sum[ Sum[
    (-1/4/Pi*ing[[c,d]]*F[[a,c]]*F[[b,d]] +
     1/16/Pi*g[[a,b]]*F[[c,d]]*Fc[[c,d]]),
     {d,1,4}],{c,1,4}],{a,1,4},{b,1,4}]]

   nu[r]         2
  E      epsilon
{{---------------, 0, 0, 0},
          4
      8 Pi r

        lambda[r]      2                        2
    -(E          epsilon )              epsilon
  {0, ---------------------, 0, 0}, {0, 0, --------, 0},
             4                                 2
          8 Pi r                           8 Pi r

                 2          2
           epsilon  Sin[theta]
  {0, 0, 0, -------------------}}
                    2
                8 Pi r
```

It should be pointed out that the energy momentum tensor for a source-free electro-magnetic field is traceless since the Maxwell tensor—a fully antisymmetric tensor—is traceless. According to this property of the energy momentum tensor, the Ricci scalar vanishes as well. Consequently, the field equations reduce to $R = 8PiT$ where R is the Ricci tensor.

```
   Simplify[Table[Ricci[a,b,g,ing] -
   8 Pi T[[a,b]],{a,1,4},{b,1,4}] ]

      -lambda[r] + nu[r]        lambda[r]        2
  {{(E                    (-4 E          epsilon +

               3          4                 4        2
          4 r   nu'[r] - r  lambda'[r] nu'[r] + r   nu'[r]  +

               4              4
          2 r   nu''[r])) / (4 r ), 0, 0, 0},

            lambda[r]          2
          E              epsilon     lambda'[r]    lambda'[r] nu'[r]
     {0, -------------------- + ---------- + ----------------- -
                  4                 r                4
                 r

      2
    nu'[r]    nu''[r]
    ------- - -------, 0, 0},
       4         2

                                          2
               -lambda[r]    epsilon     r lambda'[r]
     {0, 0, 1 - E          - -------- + ------------ -
                                 2        lambda[r]
                                r        2 E

        r nu'[r]
     ------------, 0}, {0, 0, 0,
       lambda[r]
     2 E

                 2        lambda[r]       2     2
     (Sin[theta]   (-2 E          epsilon  - 2 r  +

            lambda[r]  2    3               3
          2 E          r  + r  lambda'[r] - r  nu'[r])) /

            lambda[r]   2
       (2 E            r  )}}
```

The solutions of these differential equations can easily be verified. With the coordinates

```
IndepVar = {t,r,theta,phi}

{t, r, theta, phi}
```

the line element is given by

```
dsrn = (1-2m/r+epsilon^2/r^2) Dt[t]^2 -
       1/(1-2m/r + epsilon^2/r^2) Dt[r]^2 -
       r^2 ( Dt[theta]^2 + Sin[theta]^2 Dt[phi]^2)

              2                         2
       Dt[r]                     epsilon    2 m         2
 -(------------------) + (1 + -------- - ---) Dt[t]  -
            2                      2       r
    epsilon    2 m                r
  1 + -------- - ---
          2       r
         r

    2            2             2            2
   r  (Dt[theta]  + Dt[phi]  Sin[theta] )
```

and the metric tensor.

```
  g = metric[dsrn,IndepVar]

              2
       epsilon    2 m
 {{1 + -------- - ---, 0, 0, 0},
           2       r
          r

                     2
                    r                              2
   {0, -(----------------------), 0, 0}, {0, 0, -r , 0},
                2            2
         epsilon  - 2 m r + r

              2          2
   {0, 0, 0, -(r  Sin[theta] )}}}
```

```
ing = Simplify[ Inverse[g] ]

             2
            r
{{---------------------, 0, 0, 0},
        2             2
  epsilon  - 2 m r + r


            2            2
     epsilon  - 2 m r + r                        -2
  {0, -(---------------------), 0, 0}, {0, 0, -r  , 0},
                 2
                r


                    2
            Csc[theta]
  {0, 0, 0, -(-----------)}}
                 2
                r
```

The two parameters can be interpreted as the charge *epsilon* of the body and the geometric mass *m*. Of course, in reality a body of considerable mass has no net charge. Therefore, the Reissner–Nordstrom solution is only of hypothetical interest. However, the Reissner–Nordstrom solution can help in the study of the more complicated Kerr solution for a rotating black hole due to the similarity of its structure.

The determinant for the Reissner–Nordstrom solution is the same as for the Schwarzschild solution. It is plotted in figure 6.5.

```
detg  =  Simplify[ Det[g] ]

    4          2
 -(r  Sin[theta] )
```

```
Plot3D[detg,{r,-2,2},{theta,-Pi,Pi},AxesLabel->{"r","theta","Det[g]"}]

  -SurfaceGraphics-
```

According to the metric of the Maxwell tensor, the energy momentum tensor reduces to

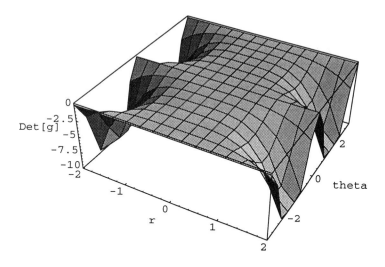

Figure 6.5.   The determinant for the Reissner–Nordstrom solution.

```
sme = { nu[r] -> - lambda[r] }

{nu[r] -> -lambda[r]}
```

```
F = F /. sme

        epsilon            epsilon
{{0, -(-------), 0, 0}, {-------, 0, 0, 0}, {0, 0, 0, 0},
         2                   2
         r                   r

  {0, 0, 0, 0}}
```

```
Fc = Fc /.sme

        epsilon            epsilon
{{0, -------, 0, 0}, {-(-------), 0, 0, 0}, {0, 0, 0, 0},
        2                   2
        r                   r

  {0, 0, 0, 0}}
```

```
T = Simplify[Table[ Sum[ Sum[
    (-1/4/Pi*ing[[c,d]]*F[[a,c]]*F[[b,d]] +
     1/16/Pi*g[[a,b]]*F[[c,d]]*Fc[[c,d]]),
    {d,1,4}],{c,1,4}],{a,1,4},{b,1,4}]]

        2         2           2
 epsilon (epsilon  - 2 m r + r )
{{-------------------------------, 0, 0, 0},
                6
          8 Pi r


                        2
                -epsilon
 {0, -----------------------------, 0, 0},
        2       2           2
    8 Pi r  (epsilon  - 2 m r + r )


          2                      2         2
   epsilon                epsilon  Sin[theta]
 {0, 0, --------, 0}, {0, 0, 0, -------------------}}
          2                           2
      8 Pi r                      8 Pi r
```

We have so far calculated all quantities sufficient to verify the field equations in a
modified form.

```
Simplify[Table[ Ricci[a,b,g,ing] - 8 Pi T[[a,b]],
{a,1,4},{b,1,4}]]

{{0, 0, 0, 0}, {0, 0, 0, 0}, {0, 0, 0, 0}, {0, 0, 0, 0}}
```

The field equations in their original forms are computed below.

```
Simplify[ Table[Einstein[a,b,g,ing] - 8 Pi T[[a,b]],
{a,1,4},{b,1,4}] ]

{{0, 0, 0, 0}, {0, 0, 0, 0}, {0, 0, 0, 0}, {0, 0, 0, 0}}
```

As a consequence the Ricci scalar obviously vanishes.

```
Simplify[RicciScalar[g,ing]]

0
```

6.6 Exercises

1. Extend the data bases in the package **PerihelionShift'** to other planets and planetary systems.

2. Find a representation of the perihelion shift using the classical parameters of an orbit. Compare your calculations to the approximations given in literature.

3. Change the package **LightBending'** in such a way that you are able to treat arbitrary masses in the calculations of light bending.

   Caution: Save the package before making changes in the program!

4. Create a three-dimensional representation of the relation for light bending (6.52) which considers changes in the mass and diameter of the star.

5. The line element in a three-dimensional space in a particular coordinate system is

$$ds^2 = dx_1^2 + x_1 dx_2^2 + x_1 \sin(x_2)^2 dx_3^2.$$

   First identify the coordinates, then examine the flatness of the metric.

6. The Minkowski line element in Minkowski coordinates

$$(x_a) = (x_0, x_1, x_2, x_3) = (t, x, y, z)$$

   is given by

$$ds^2 = dt^2 - dx^2 - dy^2 - dz^2.$$

   Is the metric flat? Determine the metric tensor.

7. Find the nonzero components of the Christoffel symbols Γ^a_{bc} of Bondi's radiating metric.

$$ds^2 = \left(\frac{V}{r} e^{2\beta} - U^2 r^2 e^{2\gamma} \right) du^2 + 2e^{2\beta} du dr +$$
$$2U r^2 e^{2\gamma} du d\vartheta - r^2 \left(e^{2\gamma} d\vartheta^2 + e^{-2\gamma} \sin(\vartheta)^2 d\phi^2 \right)$$

   where V, U, β, and γ are four arbitrary functions of the three coordinates u, r, and ϑ.

8. Verify that the Kerr form is a solution of the Einstein field equations. The Kerr form is

$$
\begin{aligned}
ds^2 \;=\; & dt^2 - dx^2 - dy^2 - dz^2 - \\
& \frac{2mr^3}{r^4 + a^2 z^2}\left(dt + \frac{r}{a^2 + r^2}(xdx + ydy) + \right. \\
& \left. \frac{a}{a^2 + r^2}(ydx - xdy) + \frac{z}{r}dz \right)^2
\end{aligned}
$$

where m and a are constants.

9. Check that the Boyer–Lindquist form of Kerr's solution is a solution of Einstein's field equations

$$
\begin{aligned}
ds^2 \;=\; & \frac{\Delta}{\rho^2}(dt - a\sin(\vartheta)^2 d\phi)^2 - \\
& \frac{\sin(\vartheta)^2}{\rho^2}\left((r^2 + a^2)d\phi - adt\right)^2 \\
& -\frac{\rho^2}{\Delta}dr^2 - \rho^2 d\vartheta^2
\end{aligned}
$$

where $\rho^2 = r^2 + a^2\cos(\vartheta)^2$ and $\Delta = r^2 - 2mr + a^2$.

7

Fractals

7.1 Measuring a borderline

A natural borderline separating two objects can be a complicated curve. When looking at a distant object governed by a geometrical structure, a skyscraper, for example, we get the impression that its borderlines are straight lines. Looking through binoculars, we observe that there are wrinkles and loops in its borderline, and a closer look reveals that the object has an even more complicated shape. Following this reasoning, we may wonder whether natural objects can be described fully by Euclidean geometry. In fact, nowhere in nature will we observe the idealized *straight line*. Nature itself uses straight lines connecting two different points only as an approximation and on small scales. Objects in our natural environment have different geometrical structures at different scales of magnification.

Let us consider a tree as an object of study. If we are far away from the tree, we can imagine that the picture we see is similar to a point or a short line on the horizon. If we get closer to the tree, the appearance changes. First, we see the extension in a plane and coming closer, we see the spatial arrangements of its branches. Up close enough, we recognize small branches and leaves. The building blocks of a tree are not geometrical objects like cylinders, balls, cones, and the like. The branches of a tree exhibit self-similarity: after scaling of a branch, a sub-branch forms from which another sub-branch can be scaled, and so on. This type of self-similar scaling law was discovered by Leonardo da Vinci, who experimented with this subject back in the 16th century [7.4]. In modern day mathematics, Benoit Mandelbrot has introduced the term *fractals* to describe such scaling laws of self-similarity.

When studying complicated natural objects, we simplify the problem by considering the three-dimensional object in a projection plane. In the case of the tree, we study the shadow of the tree in order to reduce the problem. The picture of the shadow is easily created with *Mathematica* following Gray & Glynn [7.1] (see figure 7.1). To construct the tree, simple building blocks are put together in a self-similar

Figure 7.1.   Fractal tree.

way. The package **Tree'** contains all the necessary functions to create branches
branchLine[ ] , to rotate lines **rotateLine[]**, and to scale branches **BranchScal-
ing**. A listing of the package is given below.

─────────────────────────── File tree.m ───────────────────────────

```
BeginPackage["Tree'"]

Needs["Geometry'Rotations'"];

Clear[Tree, rotateLine, branchLine, createBranches];

Tree::usage = "Tree[options___] creates a fractal tree. The options of
the function Tree determine the form of the fractal created. Options are
           Generation -> 10, \n
           BranchRotation -> 0.65, \n
           BranchSkaling -> 0.75, \n
           BranchThickness -> 0.7, \n
           OriginalThickness -> 0.07, \n
           BranchColor -> {RGBColor[0,0,0]} \n
Example: Tree[BranchColor->ll,BranchRotation->0.3] \n
         ll is a list created in the package Tree."

(* --- global variables --- *)

Generation::usage
BranchRotation::usage
BranchSkaling::usage
BranchThickness::usage
```

```
OriginalThickness::usage
BranchColorn::usage

Begin["'Private'"]

(* --- rotate a line --- *)
rotateLine[Line[{start_, end_}], angle_] :=
Line[{end, end + branchSkaling
                Rotate2D[end - start, angle Random[Real, {0.5,1.5}]
                ]}];

(* --- branch a line --- *)
branchLine[Line[points_]] := {rotateLine[Line[points],
                              branchRotation],
                              rotateLine[Line[points],
                              - branchRotation]};

(* --- change thickness --- *)
branchLine[Thickness[th_]] := Thickness[th branchThickness];

(* --- define color of a branch --- *)
branchLine[RGBColor[r_, g_, b_]] := (
                        branchColor = RotateLeft[branchColor];
                        First[branchColor] );

(* --- create branches --- *)
createBranches[lines_] := Flatten[Map[branchLine, lines]];

(* --- options if Tree[] --- *)
Options[Tree] =
     {
            Generation -> 10,
            BranchRotation -> 0.65,
            BranchSkaling -> 0.75,
            BranchThickness -> 0.7,
            OriginalThickness -> 0.07,
            BranchColor ->
                {
                   RGBColor[0,0,0]}
     };

(* --- create a tree --- *)

Tree[options___] := Block[
   {generations, branchRotation,
    branchScaling, branchThickness,
    originalThickness, branchColor},
(* --- check options --- *)
    {generations, branchRotation, branchScaling, branchThickness,
     originalThickness, branchColor} =
    {Generation, BranchRotation, BranchScaling, BranchThickness,
     OriginalThickness, BranchColor}  /. {options}
                           /. Options[Tree];
(* --- iterate the functions and display the tree --- *)
    Show[
        Graphics[
            NestList[
```

```
                    createBranches,
                    {
                            First[\alpha branchColor],
                            Thickness[originalThickness],
                            Line[{{0,0},{0,1}}]
                    },
                    generations]],
              FilterOptions[Show, options],
              AspectRatio -> Automatic,
              PlotRange -> All]];

(* --- filter for options --- *)

FilterOptions[ command_Symbol, opts___] :=
     Block[{keywords = First /@ Options[command]},
          Sequence @@ Select [{opts},
                            MemberQ[keywords, First[#]]&]]
End[]
EndPackage[]

(* --- an example of a color list --- *)


ll = {RGBColor[0.5620000000000001, 0.236, 0.071],
       RGBColor[0.5470000000000001, 0.229, 0.06900000000000001],
       RGBColor[0.5, 0.21, 0.063], RGBColor[0.469, 0.196, 0.059],
       RGBColor[0.033, 0.281, 0.035], RGBColor[0.046, 0.395, 0.05],
       RGBColor[0.055, 0.469, 0.059],
       RGBColor[0.07000000000000001, 0.602, 0.076],
       RGBColor[0.085, 0.727, 0.092], RGBColor[0.109, 0.937, 0.118],
       RGBColor[0.013, 0.75, 0.028]}
```

One of the characteristic properties of a projected tree is the length of its boundary line. If we choose a fixed yardstick length for determining the length of the boundary line, we get its total length by the number of yardsticks multiplied by the length of the yardstick. The mathematical formula is $L = N(\epsilon)\epsilon$, where L is the resulting length, ϵ is the length of the yardstick, and $N(\epsilon)$ the number of yardsticks used to cover the boundary.

In a second experiment we change the length of the yardstick ϵ. We again count a number $N(\epsilon)$ and calculate length L by the same formula as above. The first observation we make is that the calculated length L has a different value compared to the first measurement. For example, if we choose the yardstick length measuring our tree to be the vertical height of the tree, we get a different length compared to measuring the tree with a small yardstick of about 1 cm. The first measurement of the boundary line is a very crude estimation of its actual length. The accuracy of the measurement increases with the decrease in length of the yardstick used. Not only does the accuracy of the measurement increase, the numerical value of the total length L increases as well. The method of measuring the length of the boundary line by means of a yardstick is called the yardstick method.

Another method for determining the length of a boundary line is the box counting method. In this method, the object is superimposed on a lattice with mesh-size ϵ. If we count the squares which contain a part of the boundary and multiply the number of boxes $N(\epsilon)$ by mesh-size ϵ, we get an approximated length of the boundary line. Again we observe that with decreasing mesh-size ϵ, the accuracy of the measurement and the total length L increases. The number of boxes counted in the box counting method is nearly of the same order as the number of yardsticks in the yardstick method.

If length L increases while yardstick ϵ decreases, the question arises whether there exists a finite length of the boundary of the tree? If the length of the boundary is finite, we expect that the number of yardsticks $N(\epsilon)$ must increase proportionally to $1/\epsilon$ (i.e., $N(\epsilon) = L_N/\epsilon$). In other words, if the length of the boundary is $L = N(\epsilon)\epsilon = L_N$ where L_N is a constant for any $\epsilon \to 0$, we can say that the length is constant. If we apply this thought experiment to a real, natural object and count the number of boxes, we observe a completely different behavior.

The measurement of natural objects like blood cells or the bronchial tree using the yardstick or box counting method shows a different relationship between the yardstick length and the number $N(\epsilon)$. The actual relation observed in experiments [7.2, 7.3] is $N(\epsilon) = a\epsilon^{-D}$ where D is a number greater than one for plane objects. If we insert the experimentally observed relation for the number of yardsticks into the length relation $L = N\epsilon$, we get

$$L(\epsilon) = N(\epsilon)\epsilon = a\epsilon^{1-D}. \qquad (7.1)$$

This relation applies to any boundary line. For a Euclidean curve which is smooth and differentiable at any point, we expect that parameter a represents the finite length L_N and that dimension D equals 1 as $\epsilon \to 0$. For natural objects the dimension D is equal to one. The property that the dimension of a natural object is different from its topological dimension was used by Mandelbrot to define the term 'fractal' [7.4]. The experimental determination of dimension D follows from the slope of a log-log plot in which the length of the curve is plotted versus the length of the yardstick. The slope of the plot is equal to $1 - D$. In fractal theory the quantity

$$D = 1 + \frac{\log(L(\epsilon))}{\log(1/\epsilon)} - \frac{\log(a)}{\log(1/\epsilon)} \qquad (7.2)$$

is called the fractal dimension. This parameter characterizes the plane filling of the curve. The tree example used earlier in this chapter is illustrative for our purposes but too complicated to determine the fractal dimension by analytical methods. Another example of a fractal object is the curve as defined by Koch, who at the turn of the century introduced the mathematical *monster* known as the Koch snowflake. At the same time other mathematicians including Cantor, Peano, and Weierstrass discussed sets of points and curves with very strange properties. An example of this type of curve is given in figure 7.2, which shows the Koch snowflake. Using the Koch curve, we can show how the fractal dimension of such a

Figure 7.2.   Koch's snow flake.

curve (which is nowhere differentiable) is determined and how self-similarity occurs.

7.2 The Koch curve

We have been discussing self-similarity, especially of self-similar curves, but have not explained what is meant by a self-similar object. An example of a self-similar object from geometry is the congruent triangle. Everybody knows that the theorem of Pythagoras $c^2 = a^2 + b^2$ is satisfied for a right triangle. In this formula, c denotes the hypotenuse and a and b the legs of a right triangle (see figure 7.3). The proof of the Pythagorean theorem is given by the self-similar properties of the triangle.

The area of a right triangle is determined by the length of the hypotenuse and the smaller of the two angles between the hypotenuse and its legs ϕ; i.e., $F = f(c, \phi)$. Since F has the dimension of area and c has the dimension of length, we can write $F = c^2\Phi(\phi)$. Drawing the normal line of the hypotenuse through the right angle, we divide the total triangle into two self-similar triangles (see figure 7.3). The areas of the self-similar triangles are $F_1 = a^2\Phi(\phi)$ and $F_2 = b^2\Phi(\phi)$ where $\Phi(\phi)$ is the same function for both (similar) triangles. The sum of the areas F_1 and F_2 is the total area F of the triangle

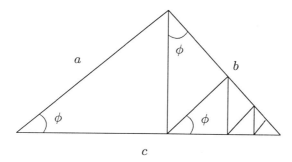

Figure 7.3. Self-similarity on a rectangular triangle.

$$F \quad = \quad F_1 + F_2 \tag{7.3}$$

$$c^2 \Phi(\phi) \quad = \quad a^2 \Phi(\phi) + b^2 \Phi(\phi). \tag{7.4}$$

Cancelation of the mutual function yields

$$c^2 = a^2 + b^2 \tag{7.5}$$

Q.E.D.

This sort of self-similarity is known as congruence in geometry. If we apply this construction again to divide the right triangle for each triangle and repeat the procedure ad infinitum, we get a sequence of triangles which are scaled versions of the original triangle. At each level of division we find the same triangles, but scaled by a different factor. This behavior of repetition and scaling was used by Helge von Koch to construct the Koch curve.

The initial element of the Koch curve is a straight line of length $L_N = 1$. The first step in constructing the Koch curve is a scaling of the total length by a factor $r = 1/3$. In the second step, four elements are arranged as shown in figure 7.4. From this figure we see that the curve loses its differentiability at the connection points of the four lines. These two fundamental steps can be infinitely applied to each of the line elements. In a k-th iteration step we get a total scaling factor of $r_k = (1/3)^k$. The number of line elements increases up to $N_k = 4^k$. The first three steps of this construction are shown in figures 7.4–7.6. If we measure the length of the Koch curve by a yardstick of the same length as the scaling factor $\epsilon = r$, we find the equation from the length relation $L(\epsilon) = N(\epsilon)\epsilon$

$$D = \frac{\log N}{\log(1/\epsilon)}, \tag{7.6}$$

and one obtains $D = \log 4/\log 3 = 1.218\ldots$ for the Koch curve. Thus the fractal dimension for a self-similar curve follows from the number of building blocks N of

Figure 7.4.   First iteration of the Koch curve.

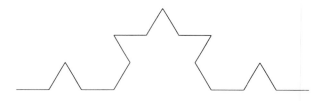

Figure 7.5.   Second iteration of the Koch curve.

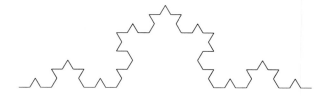

Figure 7.6.   Third iteration of the Koch curve.

the generator and the scaling factor r which is used as the yardstick length. The geometrical structure of the line elements is not contained in the fractal dimension, because the fractal dimension is not a unique property of a curve. Thus we get the same fractal dimension for curves with completely different appearances (compare figures 7.6 and 7.7).

The Koch curve is constructed in *Mathematica* with the function **Line[ ]**. We define the generator of the Koch curve in the **Koch[ ]** function which is part of the **Koch'** package, and use the *Mathematica* function **Map[ ]** to generate the higher iterations of the generator. By keeping the generator and the iteration separate in the creation process of the fractal curve, we are able to mix two or more generators into the iteration process. In figure 7.8, the Koch generator is mixed with a rectangular representation. The first two iterations are done with the original Koch generator. The next two iterations use the rectangular Koch generator. Separating the iteration process from the definition of the fundamental generators allows any mixing of generators in any state of the iteration. In the package **Koch'** we define a number of generators of fractal curves. Their combinations are accessed by the function **Fractal[ ]**. This function uses a string containing one of the possible fractals as the first argument. The second argument of the function changes the default values of the generators.

Another form of the Koch curve is obtained if we change the base angle α of the triangle in the generator. If we again use four line elements to set up the generator and alter the scaling factor to $r = 1/(2 + 2\cos(\alpha))$, we find a fractal dimension of

$$D = \frac{\log 4}{\log 2 + \log(1 + \cos(\alpha))}. \tag{7.7}$$

A representation of the dimension D versus the angle α is given in figure 7.9. In the case of $\alpha = 0$, the dimension is reduced to $D = 1$, and for $\alpha = \pi/2$, the maximum dimension $D = 2$ occurs. For $D = 2$, we have a plane filling curve. For the specific value $\alpha = 1.4$, the 6th iteration of the Koch curve with a variable base angle is given in figure 7.10.

──────────────────── File koch.m ────────────────────

```
BeginPackage["Koch'"]

Clear[Koch,VKoch,WKoch,QKoch,Quad,NGon,docurve,Fractal,FaktalPlot]

Needs["Geometry'Rotations'"]

Fractal::usage = "Fractal[curve_String, options___] creates a graphical
representation of a fractal curve. The type of curve is determined by
the first argument. A list of available curves is obtained by calling
Fractal[List] or Fractal[Help]. The second argument allows to change the
options of the function. The default values are Generations -> 3,
Angle -> Pi/6 and Corners -> 6."

Generations::usage
Angle::usage
```

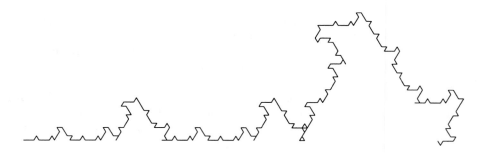

Figure 7.7.   Fourth iteration of an altered Koch curve. The triangle is located at the right end of the unit base element.

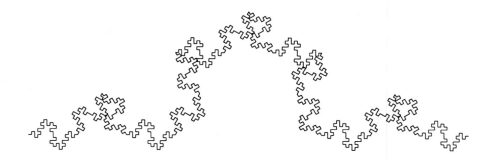

Figure 7.8.   Mixing of two generators. The first two iteration steps are governed by the original Koch generator. In the last two iteration steps a rectangular Koch generator is used.

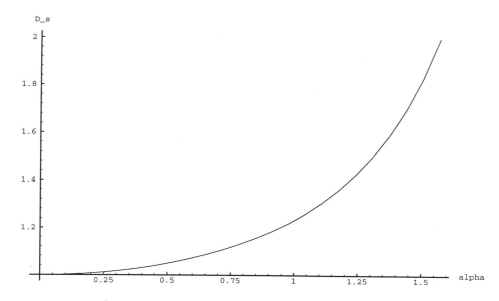

Figure 7.9. Change of the fractal dimension with a change of base angle.

Figure 7.10. Koch curve with base angle $\alpha = 1.4$. The scaling factor is $r = 0.42736\ldots$.

```
Corners::usage

Begin["'Private'"]

(* --- generator of the Koch curve --- *)
(*              __/\__                  *)

Koch[Line[{StartingPoint_,EndPoint_}]]:=Block[{fActor, angle, liste={}},
      fActor = 1/3;
      angle  = Pi/3;
      l1 = StartingPoint;
      l2 = StartingPoint+(EndPoint - StartingPoint)*fActor;
      AppendTo[liste,Line[{l1,l2}]];
      l1 = l2;
      l2 = l2 + Rotate2D[EndPoint-StartingPoint,-angle,{0,0}]*fActor;
      AppendTo[liste,Line[{l1,l2}]];
      l1 = l2;
      l2 = l2 + Rotate2D[EndPoint-StartingPoint,angle,{0,0}]*fActor;
      AppendTo[liste,Line[{l1,l2}]];
      l1 = l2;
      l2 = l2 + Rotate2D[EndPoint-StartingPoint,0,{0,0}]*fActor;
      AppendTo[liste,Line[{l1,l2}]]
      ]

(* --- generator of an altered Koch curve --- *)
(*              ____/\                          *)

VKoch[Line[{StartingPoint_,EndPoint_}]]:=Block[{fActor, angle, liste={}},
      fActor = 1/3;
      angle  = Pi/3;
      l1 = StartingPoint;
      l2 = StartingPoint+(EndPoint - StartingPoint)*fActor;
      AppendTo[liste,Line[{l1,l2}]];
      l1 = l2;
      l2 = l2 + Rotate2D[EndPoint-StartingPoint,0,{0,0}]*fActor;
      AppendTo[liste,Line[{l1,l2}]];
      l1 = l2;
      l2 = l2 + Rotate2D[EndPoint-StartingPoint,-angle,{0,0}]*fActor;
      AppendTo[liste,Line[{l1,l2}]];
      l1 = l2;
      l2 = l2 + Rotate2D[EndPoint-StartingPoint,angle,{0,0}]*fActor;
      AppendTo[liste,Line[{l1,l2}]]
      ]

(* --- generator of the Koch curve with variable base angle --- *)

WKoch[Line[{StartingPoint_,EndPoint_}]]:=Block[{fActor, liste={}},
      fActor = 1/(2*(1+Cos[angle]));
      l1 = StartingPoint;
      l2 = StartingPoint+(EndPoint - StartingPoint)*fActor;
      AppendTo[liste,Line[{l1,l2}]];
      l1 = l2;
      l2 = l2 + Rotate2D[EndPoint-StartingPoint,-angle,{0,0}]*fActor;
      AppendTo[liste,Line[{l1,l2}]];
      l1 = l2;
      l2 = l2 + Rotate2D[EndPoint-StartingPoint,angle,{0,0}]*fActor;
      AppendTo[liste,Line[{l1,l2}]];
```

```
        l1 = l2;
        l2 = l2 + Rotate2D[EndPoint-StartingPoint,0,{0,0}]*fActor;
        AppendTo[liste,Line[{l1,l2}]]
        ]

  (* --- generator of the rectangular Koch curve --- *)
  (*                    __
                   __|  |   __
                     |__|                          *)

  QKoch[Line[{StartingPoint_,EndPoint_}]]:=Block[{fActor, angle, liste={}},
        fActor = 1/4;
        angle  = Pi/2;
        l1 = StartingPoint;
        l2 = StartingPoint+(EndPoint - StartingPoint)*fActor;
        AppendTo[liste,Line[{l1,l2}]];
        l1 = l2;
        l2 = l2 + Rotate2D[EndPoint-StartingPoint,-angle,{0,0}]*fActor;
        AppendTo[liste,Line[{l1,l2}]];
        l1 = l2;
        l2 = l2 + Rotate2D[EndPoint-StartingPoint,0,{0,0}]*fActor;
        AppendTo[liste,Line[{l1,l2}]];
        l1 = l2;
        l2 = l2 + Rotate2D[EndPoint-StartingPoint,angle,{0,0}]*fActor;
        AppendTo[liste,Line[{l1,l2}]];
        l1 = l2;
        l2 = l2 + Rotate2D[EndPoint-StartingPoint,angle,{0,0}]*fActor;
        AppendTo[liste,Line[{l1,l2}]];
        l1 = l2;
        l2 = l2 + Rotate2D[EndPoint-StartingPoint,0,{0,0}]*fActor;
        AppendTo[liste,Line[{l1,l2}]];
        l1 = l2;
        l2 = l2 + Rotate2D[EndPoint-StartingPoint,-angle,{0,0}]*fActor;
        AppendTo[liste,Line[{l1,l2}]];
        l1 = l2;
        l2 = l2 + Rotate2D[EndPoint-StartingPoint,0,{0,0}]*fActor;
        AppendTo[liste,Line[{l1,l2}]]
        ]

  (* --- generator for a rectangular curve --- *)
  (*                    __
                   __|  |__                     *)

  Quad[Line[{StartingPoint_,EndPoint_}]]:=Block[{fActor, angle, liste={}},
        fActor = 1/3;
        angle  = Pi/2;
        l1 = StartingPoint;
        l2 = StartingPoint+(EndPoint - StartingPoint)*fActor;
        AppendTo[liste,Line[{l1,l2}]];
        l1 = l2;
        l2 = l2 + Rotate2D[EndPoint-StartingPoint,-angle,{0,0}]*fActor;
        AppendTo[liste,Line[{l1,l2}]];
        l1 = l2;
        l2 = l2 + Rotate2D[EndPoint-StartingPoint,0,{0,0}]*fActor;
        AppendTo[liste,Line[{l1,l2}]];
        l1 = l2;
        l2 = l2 + Rotate2D[EndPoint-StartingPoint,angle,{0,0}]*fActor;
```

```
                AppendTo[liste,Line[{l1,l2}]];
                l1 = l2;
                l2 = l2 + Rotate2D[EndPoint-StartingPoint,0,{0,0}]*fActor;
                (* l2 = l2 + EndPoint*fActor;*)
                AppendTo[liste,Line[{l1,l2}]]
                ]


(* --- generator for a N gon --- *)

NGon[Line[{StartingPoint_,EndPoint_}]]:=Block[{l1, l2, angle, liste={}},
        angle = 2*Pi/corners;
        l1 = StartingPoint;
        l2 = StartingPoint+(EndPoint - StartingPoint);
        AppendTo[liste,Line[{l1,l2}]];
        Do[
        l1 = l2;
        l2 = l2 + Rotate2D[EndPoint-StartingPoint,k*angle,{0,0}];
        AppendTo[liste,Line[{l1,l2}]],
        {k,1,corners-1}];
        liste
        ]


(* --- calculate the higher iterations --- *)

docurve[Type_,Linie_]:=Block[{},
        Flatten[Map[Type,Linie]]
        ]


(* --- plot of a line sequence --- *)

FractalPlot[x_]:=Show[Graphics[x],AspectRatio->Automatic]

(* --- options for Fractal[] --- *)

Options[Fractal]:={
                Generations -> 3,
                Angle -> Pi/6,
                Corners -> 6
                }

(* --- create the fractal curve --- *)

Fractal[curve_, options___]:=Block[{generations, angle, corners},
(* --- check options --- *)
        {generations,angle,corners} = {Generations,Angle,Corners}
                /. {options} /. Options[Fractal];
(* --- menu for the different fractal curves --- *)
        If[curve == "List" || curve == "Help",
        Print[" "];
        Print[" --------- available curves ---------"];
        Print["  Koch    :   Koch curve"];
        Print["  QKoch   :   rectangular Koch curve"];
        Print["  VKoch   :   altered Koch curve"];
        Print["  WKoch   :   variable angle Koch curve"];
        Print["  Quad    :   rectangular curve"];
        Print["  Star    :   Koch star"];
        Print["  Square  :   Koch square"];
```

```
          Print[" N-gon  :  Koch N gon"];
          Print[" Mixture: 2 x Koch and 2 x QKoch"];
          Print[" Random :  random generation"]];

(* --- plot the Koch curves --- *)
          If[curve == "Koch" ||
             curve == "QKoch" ||
             curve == "VKoch" ||
             curve == "WKoch" ||
             curve == "Quad",
(* --- ToExpression transforms a string to an expression --- *)
             k1 = ToExpression[curve][Line[{{0,0},{1,0}}]];
             Do[
             k1 = docurve[ToExpression[curve],k1],
             {k,1,generations}];
             FractalPlot[k1]
             ];
(* --- plot a Koch star --- *)
          If[curve == "Star",
             corners = 3;
             k1 = NGon[Line[{{0,0},{1,0}}]];
             Do[
             k1 = docurve[Koch,k1],
             {k,1,generations}];
             FractalPlot[k1]
             ];
(* --- plot a Koch square --- *)
          If[curve == "Square",
             corners = 4;
             k1 = NGon[Line[{{0,0},{1,0}}]];
             Do[
             k1 = docurve[Koch,k1],
             {k,1,generations}];
             FractalPlot[k1]
             ];
(* --- plot a Koch N gon --- *)
          If[curve == "N-gon",
             k1 = NGon[Line[{{0,0},{1,0}}]];
             Do[
             k1 = docurve[Koch,k1],
             {k,1,generations}];
             FractalPlot[k1]
             ];
(* --- plot a mixture of Koch curves --- *)
          If[curve == "Mixture",
             k1 = Koch[Line[{{0,0},{1,0}}]];
             k1 = docurve[Koch,k1];
             k1 = docurve[QKoch,k1];
             k1 = docurve[QKoch,k1];
             FractalPlot[k1]
             ];
(* --- plot a random sequence of Koch curves --- *)
          If[curve == "Random",
             listec ={Koch,QKoch,VKoch,WKoch,Quad,NGon};
             k2 = Random[Integer,{1,6}];
             k3 = Random[Integer,{1,6}];
                If[k2 == 6 || k3 == 6, corners =
```

```
                                                        Random[Integer,{3,12}]];
                name1 = listec[[k2]];
                name2 = listec[[k3]];
                k1 = name1[Line[{{0,0},{1,0}}]];
                k1 = docurve[name1,k1];
                Do[
                k1 = docurve[name2,k1],
                {k,1,generations-1}];
                FractalPlot[k1]
                ];
        ]
    End[]
    EndPackage[]
```

7.3 Multi-fractals

In the previous sections we discussed structures with mutual scaling factors. This kind of self-similarity is a special case of fractals. A more common type of fractals uses several scaling factors in competition with one another. If in the same system different scaling factors occur with different probabilities, we speak of multi-fractal behavior. The first step in the construction of a multi-fractal consists of the division of a set into j components in which each is scaled by the factor $1/r_j < 1$. We assume that each part of the j-fold set is related to a probability P_j. The probabilities P_j are normalized so that $\sum_{j=1}^{n} P_j = 1$, where n counts the number of subsets. The second step in constructing $k = 2$ is a repetition of the first step applied to each subset. The n subsets are each divided into n subsets and are related to the corresponding probabilities. A graphical representation of this division is given in figure 7.11. The multi-fractal is created as $k \to \infty$.

The consequence of this construction is that we can divide the total fractal into n parts. Each part of the fractal is scaled by a factor $1/r_j$ and the measure of the j-th part is determined by P_j. Using these quantities we can define one of the characteristic functions of a multi-fractal by

$$\chi_{q,j}(\epsilon) = \sum_{i=1}^{N} p_{j,i}^q(\epsilon) = P_j^q \chi_q(\epsilon r_j), \tag{7.8}$$

where $\chi_{q,j}(\epsilon)$ characterizes the j-th part of the fractal by a probability $p_{j,i}(\epsilon)$ ($p_{j,i}(\epsilon)$ is the i-th probability for the j-th part of the total fractal). For the total fractal we get

$$\chi_q(\epsilon) = \sum_{j=1}^{n} \chi_{q,j}(\epsilon). \tag{7.9}$$

Using relation $\chi_q(\epsilon) = \epsilon^{(q-1)D_q}$ and equation (7.8), we get the expressions

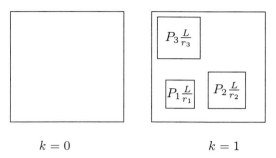

$k = 0$ $k = 1$

Figure 7.11. Representation of a multi-fractal. The initial state $k = 0$ and the first iteration $k = 1$ are shown. The scaling factors are r_1, r_2, and r_3. The related probabilities are P_1, P_2, and P_3.

$$\chi_{q,j}(\epsilon) = P_j^q \chi_q(\epsilon r_j) = P_j^q r_j^{(q-1)D_q} \epsilon^{(q-1)D_q} \tag{7.10}$$

$$\chi_q = \sum_{j=1}^{n} P_j^q r_j^{(q-1)D_q} \epsilon^{(q-1)D_q} = \epsilon^{(q-1)D_q} \tag{7.11}$$

which define the implicit equation for determining the generalized dimension D_q by

$$\sum_{j=1}^{n} P_j^q r_j^{(q-1)D_q} = 1. \tag{7.12}$$

Depending on the choice of probabilities P_j and scaling factors r_j, we can use (7.12) to derive several special cases for a multi-fractal. For $q = 0$ we get the fractal dimension $D = D_0$. This dimension was introduced by Mandelbrot [7.4] for a fractal

$$\sum_{j=1}^{n} r_j^{-D_0} = 1. \tag{7.13}$$

For arbitrary q and identical scaling factors $r_j = r$, we get the representation of D_q by

$$\sum_{j=1}^{n} P_j^q r^{(q-1)D_q} = 1$$

$$(q - 1)D_q \ln r = -\ln \sum_{j=1}^{n} P_j^q$$

$$D_q = \frac{1}{q-1} \frac{\ln\left(\sum_{j=1}^{n} P_j^q\right)}{\ln\left(1/r\right)}. \tag{7.14}$$

Once the probabilities P_j and the scaling factors r_j are equal for each individual j, the multi-fractal properties no longer occur.

Knowing the dependence of D_q on q, alternate representations of the fractal dimensions emerge. By a Legendre transformation we can introduce

$$(q-1)D_q = \tau(q) = q\alpha_q - f_q \tag{7.15}$$

where f_q is the multi-fractal distribution and α_q is the Hölder exponent. The Hölder exponent α_q is defined by the derivative of $\tau(q)$

$$\alpha_q = \frac{d}{dq}\tau(q). \tag{7.16}$$

Once we know the fractal dimensions D_q, we are able to determine the Hölder exponent and f_q by relations (7.16) and (7.15), respectively. Knowing both quantities, we can plot $f = f(\alpha)$ versus α eliminating q. Calculating the derivative of $\tau(q)$ given in (7.16) causes numerical problems. Finding the numerical derivative of the Legendre transformation of D_q is the main problem in our calculation. In the package **MultiFractal'**, we use a symmetric difference procedure (see section 5.4) for representing the numerical values of the derivatives of $\tau(q)$. The transformation to τ is defined in the function **Tau[ ]**. The approximations of derivatives by their differences result in a numerical error, but it is sufficiently small if we choose steps dq in q as a small quantity.

MultiFractal[ ] calculates the multi-fractal characteristics. Probabilities P_j and scaling factors r_j are input parameters for this function. The fractal dimension D_q, the function $\tau(q)$ and the Legendre transformation are determined by the functions **Dq[ ]**, **Tau[ ]**, and **Alpha[ ]**, respectively. After their calculation, these quantities are graphically represented by the *Mathematica* function **ListPlot[ ]**. An example of a transformation is given in figures 7.12–7.15.

```
———————————————————————— File multifra.m ————————————————————————

BeginPackage ["MultiFractal'"]

Clear[Dq, Tau, Alpha, MultiFractal]

MultiFractal::usage = "MultiFractal[p_List,r_List] calculates the
multifractal spectrum D_q for a model based on the probabilities p and the
scaling factors r. This function plots five functions Tau(q), D_q(q),
Alpha(q), f(q) and f(Alpha)."

Begin["'Private'"]

(* --- calculate the multifractal dimensions --- *)

Dq[p_List, r_List] := Block[{l1, l2, listrg = {}},
(* --- length of the lists --- *)
```

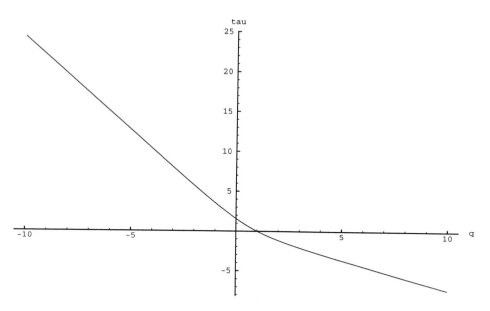

Figure 7.12. Function $\tau_q = (q-1)D_q$ versus q in the range $q \in [-10, 10]$ for the model fixed by $n = 3$ and $r = 1/2$. The probabilities are $P_1 = 1/5$, $P_2 = 3/5$ and $P_3 = 1/5$.

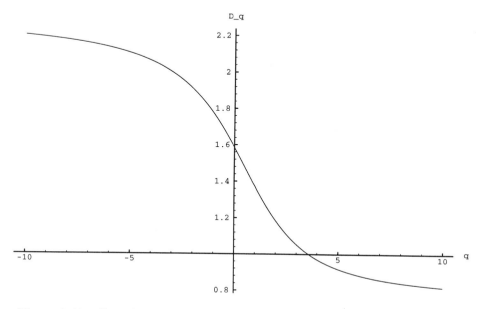

Figure 7.13. Generalized fractal dimension D_q for the model $n = 3$, $r = 1/2$, $P_1 = 1/5$, $P_2 = 3/5$ and $P_3 = 1/5$ $q \in [-10, 10]$.

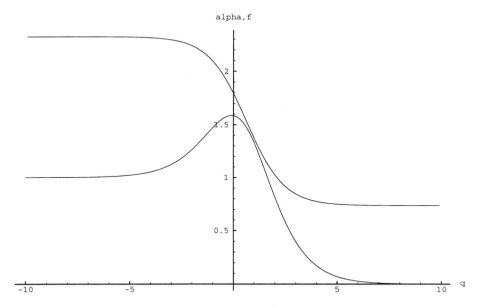

Figure 7.14. The exponent α_q (top) and f_q (bottom) versus q for the model $n = 3$, $r = 1/2$, $P_1 = P_3 = 1/5$, $P_2 = 3/5$, and $q \in [-10, 10]$.

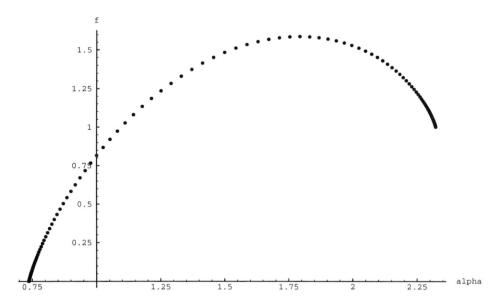

Figure 7.15. The fractal spectrum $f(\alpha)$ for a multi-fractal with $n = 3$, $r = 1/2$, $P_1 = P_3 = 1/5$, $P_2 = 3/5$.

```
    l1 = Length[p];
    l2 = Length[r];
    If[l1 == l2,
(* --- variation of q and determination of D_q --- *)
      Do[
      gl1 = Sum[p[[j]]^q r[[j]]^((q - 1) Dfractal), {j, 1, l1}] - 1;
      result = FindRoot[gl1 == 0, {Dfractal, -3, 3}];
      result = -Dfractal /. result;
(* --- collect the results in a list --- *)
      AppendTo[listrg,{q,result}],
      {q,-10,10,.1}],
    Print[" "];
    Print["  Lengths of lists are different!"];
    listrg = {}
    ];
      listrg
      ]


(* --- calculate Tau --- *)

Tau[result_List]:=Block[{l1, listtau = {}},
(* --- lengths of the lists --- *)
    l1 = Length[result];
(* --- calculate Tau --- *)
    Do[
      AppendTo[listtau,{result[[k,1]],result[[k,2]] (1-result[[k,1]])}],
    {k,1,l1}];
    listtau
    ]


(* --- Legendre transform --- *)

Alpha[result_List]:=Block[{l1, dq, listalpha = {},listf = {},
                      listleg={}, mlist = {}, pl1, pl2},
(* --- lengths of the lists --- *)
    l1 = Length[result];
(* --- determine the differential dq --- *)
    dq = (result[[2,1]]-result[[1,1]]) 2;
(* --- calculate Alpha by numerical differentiation --- *)
    Do[
    AppendTo[listalpha,{result[[k,1]],(result[[k+1,2]]-result[[k-1,2]])/dq}],
    {k,2,l1-1}];
    l2 = Length[listalpha];
(* --- calculate f and collect the results in a list --- *)
    Do[
      AppendTo[listf,{result[[k,1]],
                      -(result[[k,1]] listalpha[[k,2]]-result[[k,2]])}];
      listalpha[[k,2]] = -listalpha[[k,2]],
      {k,1,l2}];
(* --- list of the Legendre transforms --- *)
          Do[
            AppendTo[listleg,{listalpha[[k,2]],listf[[k,2]]}];
            AppendTo[mlist,listf[[k,2]]],
          {k,1,l2}];
(* --- plot f and alpha versus q --- *)
      pl1 = ListPlot[listalpha,PlotJoined->True,
                  AxesLabel->{"q","alpha"},
```

```
                          Prolog->Thickness[0.001]];
         pl2 = ListPlot[listf,PlotJoined->True,AxesLabel->{"q","f"},
                          Prolog->Thickness[0.001]];
         Show[{pl1,pl2},AxesLabel->{"q","alpha,f"}];
(* --- plot the Legendre transform f versus alpha --- *)
         ListPlot[listleg,AxesLabel->{"alpha","f"}];
(* --- print the maximum of f = D_0 --- *)
         maxi = Max[mlist];
         Print[" "];
         Print["   D_0 = ",maxi]
         ]


(* --- calculate the multifractal properties --- *)

MultiFractal[p_List,r_List]:=Block[{listDq, listTau},
(* --- determine D_q --- *)
     listDq = Dq[p,r];
     ListPlot[listDq,PlotJoined->True,AxesLabel->{"q","D_q"},
              Prolog->Thickness[0.001]];
(* --- calculate Tau --- *)
     listTau = Tau[listDq];
     ListPlot[listTau,PlotJoined->True,AxesLabel->{"q","tau"},
              Prolog->Thickness[0.001]];
(* --- determine the Hoelder exponent --- *)
     Alpha[listTau]
     ]
End[]
EndPackage[]
```

7.3.1   Multi-fractals with common scaling factor

We now consider a multi-fractal with a fixed and a mutual scaling factor $r_i = r$. To determine the generalized dimensions D_q, we use equation (7.14) which gives

$$D_q = \frac{1}{q-1} \frac{\ln \left(\sum_{j=1}^{n} P_j^q \right)}{\ln(1/r)}. \tag{7.17}$$

In the following we consider a model which contains three independent sets, $n = 3$, characterized by the probabilities $P_1 = 1/5$, $P_2 = 3P_1$, and $P_3 = P_1$. If we use relation (7.14) for these three processes, we get

$$D_q = \frac{1}{q-1} \frac{\ln \left(P_1^q + P_2^q + P_3^q \right)}{\ln(1/r)}. \tag{7.18}$$

Normalizing the probability by $P_3 = 1 - P_1 - P_2$ simplifies expression (7.18) to

$$D_q = \frac{1}{q-1} \frac{\ln \left(P_1^q + P_2^q + (1 - P_1 - P_2)^q \right)}{\ln(1/r)}. \tag{7.19}$$

The numerical results are represented in figure 7.13 which is created by **MultiFractal[{1/5,3/5,1/5},{1/2,1/2,1/2}]**. In the above case, probabilities $P_1 = P_3$ and $P_2 = 3P_1$ simplify equation (7.19) to

$$D_q = \frac{1}{q-1} \frac{\ln(2+3^q) + q \ln P_1}{\ln(1/r)}. \tag{7.20}$$

From relation (7.20), we can derive analytic relations for the Hölder exponent α_q, and for the spectrum f_q by using relations (7.16) and (7.15). We get for α_q the expression

$$\alpha_q = \frac{1}{\ln(1/r)} \left\{ \frac{3^q \ln 3}{2 + 3^q} + \ln P_1 \right\}. \tag{7.21}$$

The spectrum of the fractal dimensions is given by

$$f_q = \frac{1}{\ln(1/r)} \left\{ q \frac{d}{dq} \ln(2 + 3^q) - \ln(2 + 3^q) \right\}. \tag{7.22}$$

Relation (7.22) is independent of P_1 and only contains the ratios of the probabilities. Since the expressions for D_q, α_q and f_q can not be solved explicitly, we use the numerical method implemented in the function **MultiFractal[ ]** to find the solution. Figures 7.12–7.15 show the results of our calculation. The fractal dimension D_0 of our model is $D_0 = 1.58\ldots$.

Figure 7.12 represents the auxiliary function $\tau_q = (q-1)D_q$ which is the basis of the numerical calculations. Figure 7.13 contains the representation of the generalized dimension D_q. Relations (7.16) and (7.15) for f_q and α_q are shown in figure 7.14. We observe that α_q is a monotonically decreasing function and that f_q shows its maximum at $q = 0$. The Legendre transform of these relations results in the function $f(\alpha)$ as shown in figure 7.15. We observe that the values of $f(\alpha)$ are almost equally spaced at the maximum and become denser at the boundaries of the α interval. In the $\alpha_{-\infty}$ limit, the function $f(\alpha)$ tends to 0 but for $\alpha_{+\infty}$ a finite value $f(\alpha)$ results. This means that for $\alpha = \alpha_{+\infty}$ a finite dimension of the subsets exists which is smaller than D_0 but greater than zero.

7.4 The renormalization group

Renormalization group theory is useful for describing physical phenomena that show the same behavior on different scales. We assume that p is a quantity measured with a certain accuracy. The same physical quantity is measured in a second experiment yielding p' with an accuracy which is smaller by a factor of two than the first measurement. We assume there is a resolution transformation f_2 connecting the two measurements by

$$p' = f_2(p),$$

where subscript 2 denotes the order of resolution. If we decrease the resolution of the measurement by another factor of two, we get the relation

$$p'' = f_2(p') = f_2(f_2(p)) = f_2 \cdot f_2(p) = f_4(p).$$

The general representation of our resolution transformation for two arbitrary resolutions a and b is given by

$$
\begin{aligned}
f_a \cdot f_b &= f_{ab} \\
f_1 &= 1,
\end{aligned}
$$

where 1 represents the identity transform. Applying the resolution transformation to any physical state, a reduced state containing less information is created. Decreasing the resolution from a state with small resolution is, in general, not possible. In other words, the function f cannot be inverted in general. A set of functions which is not $1-1$ is called a semi-group in mathematics. In physics the transformation reducing the resolution is called renormalization. (Strictly speaking, f should be called a semi renormalization group.) By definition, the renormalization group is closely related to the definition of a fractal.

Since a fractal stays invariant under a scaling transformation it is evident that a fractal also stays invariant under a renormalization transformation. Chronologically both terms—fractal and renormalization—were introduced in the 1970s. Both describe the behavior of an object with changing scales. The difference between the two terms is that a fractal is based on geometrical properties whereas renormalization considers the physical properties in a scaling process. However, recent developments in fractal theory also consider physical properties while renormalization theory is also applied to geometric objects. Consequently the distinction between a fractal and renormalization theory is disappearing.

Renormalization theory is a tool describing critical phenomena like phase transitions in a liquid. Liquids, for example, possess a critical point in their phase diagrams. Renormalization theory is used to describe the behavior of the system in the immediate neighborhood of the critical point. Let us consider a state of liquid below the critical point where a mixture of liquid and gas coexists. Below the critical point the mixture contains more liquid than gas. If we "coarse grain" our observation, we get a system which is dominated by the liquid phase. The combination of cells containing liquid and gas components produces a liquid state under renormalized conditions. The repetition of the "coarse graining" process results in a global liquid state. If, on the other hand, the initial state of the phase diagram contains more gas than liquid, the renormalization results in a gaseous state.

In another example we consider the renormalization procedure in connection with percolation theory. Percolation theory is a theory describing the connections in a network of random links. The theoretical basis for this theory was created by P.G. de Gennes [7.7], winner of the 1991 Nobel Prize. He applied percolation theory for disordered materials in polymer science. Percolation phenomena are widespread in nature, occuring in biological, chemical and physical systems.

Percolation theory allows the connection of two different boundaries with a cluster of particles on a lattice. Specifically, let us examine the transport of electrons through a porous medium which is located between two metal plates. The transport of the charge is carried by a percolation cluster connecting both plates. In order

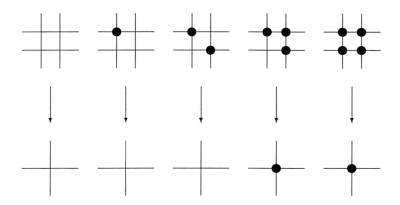

Figure 7.16.   Renormalization steps with 2×2 blocks.

to study the transport of electrons, picture the simulation of a current in a porous medium on a two-dimensional lattice. Atoms carry the charge on the lattice. The atoms are randomly scattered. Using the probability p, an atom at a certain location on the lattice can be located.

The renormalization step on this lattice is defined by the rule valid for a 2×2 sub-lattice, which is called the virtual lattice. We are able to replace the region of the virtual lattice with a new lattice point in the renormalized lattice. The resultant lattice is called the super lattice. The 2×2 cells of the virtual lattice are called blocks (see figure 7.16).

The transition from the original lattice to the super lattice follows rules for replacing old atoms wth new ones. The simplest rule applies if we have 4 atoms in a block. In this case the new point in the super lattice is an atom. If we only have three atoms in a block, another new atom emerges on the super lattice. Accordingly percolation clusters can form horizontally as well as vertically. If a block only contains one or two particles, it is impossible for a percolation cluster to occur which is independent of any direction. Therefore, no atom appears on the super lattice. Applying the transition rules as defined in a probability projection, we can write down the probability for finding an atom on the super lattice by

$$p' = f_2(p) = p^4 + 4p^3(1 - p). \qquad (7.23)$$

The first term describes the probability that all four atoms are present in a block. The second term takes into account the four possible arrangements of three atoms in a block. Since we now know the function f_2 we can determine the phase transition by using the properties of f_2.

Generalizing relation (7.23) for a lattice with $n = b \times b$ locations on which m empty points exist is given by the expression

$$f(p)_m^n = \sum_{i=0}^{m} \binom{n}{i} p^{(n-i)}(1-p)^i. \tag{7.24}$$

Equation (7.24) specifies the probability on a lattice if the block contains n locations of which all m points are empty.

The critical point p_c on the 2×2 lattice is defined in such a way that the probability will not change under the transformation f_2. The fixed point p_c is derived from the relation

$$p_c = p_c^4 + 4p_c^3(1-p_c) \tag{7.25}$$

with solutions

$$p_c = \left\{0, 1, \frac{1 \pm \sqrt{13}}{6}\right\}. \tag{7.26}$$

The numerical values of the third and fourth solutions are -0.434, and 0.768. Since p is a probability which is always greater than zero, we have to exclude the solution $p_c = -0.434$ from the physical solution set. The cases $p_c = 0$ and $p_c = 1$ are trivial since they correspond to an empty or occupied lattice. The remaining value of $p_c = 0.768$ seems to be the critical value for which a percolation takes place. We observe a gap if we compare the theoretical value with the value $p_c = 0.59$ yielded by computer simulations. However, the experimentally determined value of $p_c = 0.752$ is fairly close to its theoretical counterpart [7.5, 7.6]. A graphical representation of the critical probability versus the number of lattice points is given in figure 7.17. The curves in this figure represent different super lattices.

To see how other solutions of (7.26) are reached, we first consider the case $p < p_c$. In this case we get the inequalities

$$p_c > p > f_2(p) > f_{2^2}(p) \cdots > f_{2^n}(p). \tag{7.27}$$

Relation (7.27) shows that the probability p decreases in each renormalization step. After infinitely many renormalization steps we get the limit $f_\infty(p) = 0$. In other words, a point with an atom somewhere on the lattice is impossible, since the lattice is empty.

For the case $p > p_c$, the reverse occurs and $f_\infty(p) = 1$. After infinitely many renormalization steps, the super lattice is fully occupied. This means that all initial values in the neighborhood of $p_c = 0.768$ will tend to be $p_c = 0$ or $p_c = 1$. The fixed point at $p_c = 0.768$ is unstable (see figure 7.18).

In the following we determine the fractal dimension of the cluster at percolation $p_c = 0.768$. If an atom is present on the super lattice, we know that there are either three or four atoms in a block. The expectation value $p_c N_c$ of occupied lattice points is thus given by

$$p_c N_c = 4p_c^4 + 3 \cdot 4p_c^3(1-p_c) \tag{7.28}$$

$$\Longleftrightarrow \quad N_c = 4p_c^3 + 3 \cdot 4p_c^2(1-p_c) \tag{7.29}$$

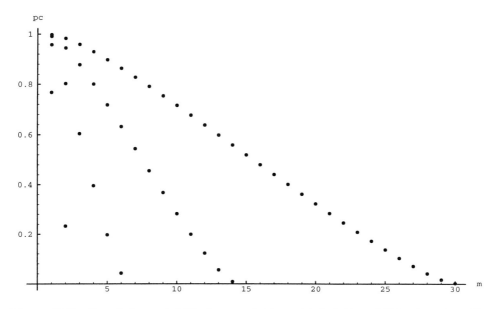

Figure 7.17.   Percolation probability for super lattices with 4, 8, 16, and 32 lattice points. The probability is plotted versus the number of empty lattice points.

where N_c is the mean value of atoms provided that the super lattice is occupied. The general formula for a square grid has the representation

$$N_c(p)_m^n = \sum_{i=0}^{m} \binom{n}{i} (n-i) p^{(n-i-1)} (1-p)^i. \qquad (7.30)$$

Equation (7.30) counts the mean number of occupied lattice points for a square lattice with n locations and with m empty locations. A graphical representation of N_c versus m is given in figure 7.19. The curves in the figure represent different block sizes.

The mesh-size in the super lattice is twice that of the original lattice. If we divide the mesh-size by 2 in the super lattice we observe N_c atoms, the average

Figure 7.18.   Stability of the fixed points in the renormalization procedure.

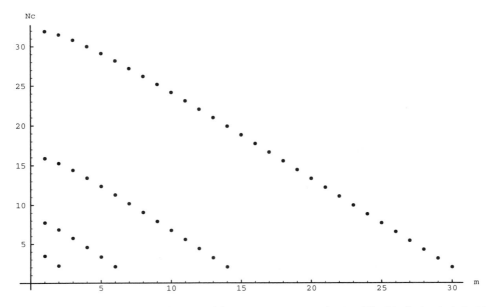

Figure 7.19.   Mean number of occupied locations in a square lattice. The block size is 4, 8, 16
and 32 as shown in the curves from bottom to top.

in the original lattice. Generalizing this observation when reducing the observation
scale by a factor of $1/b$ yields

$$N_c(b) = b^{-D}. \tag{7.31}$$

In the example discussed above $b = 2$. From relation (7.31) we get for the specific
case

$$D = \frac{\ln N_c}{\ln 2} = 1.79 \tag{7.32}$$

where constant D represents the fractal dimension of the percolation cluster. $D =$
1.79 is in good agreement with the value found in computer simulations. However,
the experimental value of the fractal dimension is different ($D = 1.9$ [7.5]). Figure
7.20 represents the fractal dimension compared to the empty lattice points for
several block sizes. We observe from this figure that the fractal dimension decreases
with an increase of empty lattice points. The dimension D approaches 2 if the
lattice is almost fully occupied.

In our previous considerations, we calculated the fractal cluster dimension at
the critical point. Other interesting quantities in the neighborhood of the critical
point are the critical exponents. The critical exponents are easy to derive if we
again use the renormalization procedure. As an example, we determine the critical
exponent of the correlation length.

For $p < p_c$ and p in the neighborhood of p_c we can represent the correlation
length ξ by

$$\xi = \xi_0|p_c - p|^{-\nu} \tag{7.33}$$

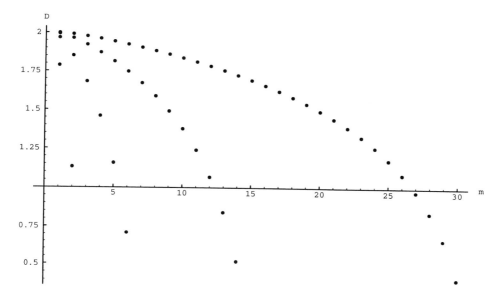

Figure 7.20. Fractal dimension of a percolation cluster versus empty locations for four block sizes 4, 8, 16 and 32.

where ξ_0 is a characteristic length of the system (e.g., the mesh-size). If we consider the rescaled super lattice, we find for the invariant correlation length

$$\xi = \xi_0' |p_c - p'|^{-\nu} \tag{7.34}$$

with $\xi_0' = 2\xi_0$. From equations (7.32) and (7.33), we derive the critical exponent ν

$$\nu = \frac{\log 2}{\log \left\{ \dfrac{p_c - p'}{p_c - p} \right\}}. \tag{7.35}$$

At the limit where p and p' tend to p_c, we can replace

$$\frac{p_c - p'}{p_c - p} \longrightarrow \frac{\partial p'}{\partial p} \Big|_{p=p_c}. \tag{7.36}$$

The final result for the critical exponent is

$$\nu = \frac{\log 2}{\log \dfrac{\partial f_2(p)}{\partial p} \Big|_{p=p_c}}. \tag{7.37}$$

Using the functional relation f_2 in equation (7.37), the numerical value $\nu = 1.4$ is close to the experimental value of $\nu = 1.35$.

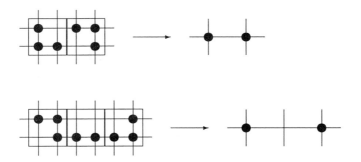

Figure 7.21.   Errors in the renormalization of a 2×2 lattice.

The renormalization group theory is useful for determining fractal and critical properties of a system. Note that the renormalization theory is a kind of perturbation theory. Errors occur in the renormalization procedure when defining renormalization rules. For example, blocks containing more than two atoms are replaced by atoms on the super lattice whereas blocks containing one or two atoms are given by void. This coarse graining process is the source of renormalization errors; i.e., we create a crude picture of the original lattice in the super lattice containing links and gaps on sites where no links were present in the original lattice (see figure 7.21). To minimize errors, we use large block sizes. If we use blocks of size b we have b^2 lattice points. The number of states in the block is given by 2^{b^2} and increases rapidly with block size b. From a practical point of view, $b = 4$ is the upper limit for which we can calculate the renormalized function f_b.

The package **Renormalization'** contains the functions **Nc[ ]** for determining the mean number of occupied lattice points; **Dim[ ]** for calculating the fractal dimension; and **Pcrit[ ]** for calculating the critical probability of percolation. Function **RenormPlot[ ]** allows the graphical representation of the above functions. Examples of the plots are given in figures 7.17, 7.19 and 7.20.

```
─────────────────────────────── File renormal.m ───────────────────────────────

BeginPackage["Renormalization'"]

Clear[f, Pcrit, Nc, Dim, RenormPlot];

Nc::usage = "Nc[n_] determines the mean number of atoms at the probability
p_c if m is changed in the range 1 <= m <= n-2. The size of the block is
determined by n."

Dim::usage = "Dim[n_] calculates the fractal dimension for the critical
probability p_c. The dimension depends on m where 1 <= m <= n-2, n is
the size of the block used."
```

```
Pcrit::usage = "Pcrit[n_] determines the critical probability for an n x n
grid under the variation of m where m is the number of empty locations in the
grid. The range of m is 1 <= m <= n-2."

RenormPlot::usage = "RenormPlot[n_,typ_String] plots the functions Nc, Dim
or Pcrit."

Begin["'Private'"]

(* --- auxilary function --- *)

f[p_,n_,m_]:= Sum[Binomial[n,i]*p^(n-i)*(1-p)^i,{i,0,m}]

(* --- mean number of particles on a grid --- *)

Nc[n_]:= Block[{p, ncliste={}},
            p = Pcrit[n];
            Do[
                AppendTo[ncliste,
                Sum[Binomial[n,i]*(n-i)*p[[k]]^(n-i-1)*(1-p[[k]])^i,
                    {i,0,k}]],
            {k,1,n-2}];
            ncliste
            ]
(* --- fractal dimension at the critical probability --- *)

Dim[n_]:=  N[Log[Nc[n]]/Log[Sqrt[n]]]


(* --- critical probability on a n x n grid --- *)

Pcrit[n_]:=Block[{ph, p, erg, erg1, gl1, pliste1={}},
            If[n > 2,
                Do[
                gl1 = p - f[p,n,i];
        (* --- solution of the fixpoint equation --- *)
                erg = NSolve[gl1==0,p];
                erg = p /. erg;
        (* --- use only real solutions --- *)
                    erg1 = {};
                    Do[If[Head[erg[[k]]]==Real,AppendTo[erg1,erg[[k]]]],
                    {k,1,Length[erg]}];
        (* --- looking for solutions between 0 and 1 --- *)
                    erg = Sort[erg1];
                    erg1 = {};
                    Do[If[erg[[k]] > 0.0 ,
                        AppendTo[erg1, erg[[k]] ] ],
                    {k,1,Length[erg]}];
                    ph = Min[erg1];
                    AppendTo[pliste1,ph],
                    {i,1,n-2}],
                Print["    "];
                Print[" choose n > 2 "]];
                pliste1]

(* --- plot the results --- *)
```

```
RenormPlot[n_,typ_String]:=Block[{},
     If[typ == "Pcrit",
        ListPlot[Pcrit[n],AxesLabel->{"m","pc"}],
        If[typ == "Nc",
           ListPlot[Nc[n],AxesLabel->{"m","Nc"}],
           If[typ == "Dim",
              ListPlot[Dim[n],AxesLabel->{"m","D"}],
              Print[" "];
              Print["  Wrong key word use: "];
              Print["  Pcrit, Nc or Dim.   "];
              Print[" "]
           ]
        ]
     ]
  ]
End[]
EndPackage[]
```

7.5 Exercises

1. Use the **Tree[ ]** function to create different kinds of trees. Which option determines the shape of a tree?

2. Extend the **Koch[ ]** function to other generators; e.g., the Peano curve. For a set of generators consult the book by Mandelbrot [7.4].

3. Examine the multi-fractal properties of a system with different numbers of probabilities and scaling factors. Determine the fractal dimensions D_0 and D_1 and give a graphical representation of these dimensions versus the number of scaling factors.

4. Use hexagonal lattices in the renormalization procedure for the percolation model.

A

Appendix

This appendix contains some information on the installation of the accompanying software and a short description of the functions defined in the packages. It also summarizes the *Mathematica* functions used in the book.

A.1 Program installation

The book is accompanied by a disk containing a collection of *Mathematica* programs and notebooks. The disk is readable by all MS-DOS-compatible PCs. The names of the programs have been chosen in such a way that they are identical with the names of the *Mathematica* packages. The name of the programs appears centered on the line indicating a new *Mathematica* section (see the text). The names of the files are reduced to a length of 8 characters. To use the programs, copy their entire content from the disk to your computer. Create a separate directory in which to store the files. If you start *Mathematica* from this directory and set the path by **SetDirectory[ ]** it is easy to read any packages by ≪ **name**.

The disk contains ASCII files so it is easy to transfer the files via *ftp* or other means to another type of computer system. The programs were developed using *Mathematica* on a PC and tested on SUN, HP and SGI computers.

A.2 Glossary of files and functions

This section contains a short description of all functions defined in the packages of this book. The packages are alphabetically listed.

anharmon.m An harmonic oscillator of quantum mechanics.

AsymptoticPT AsymptoticPT[N_,kin_] determines the asymptotic approximation for $|x| \to \infty$ for the continuous case of eigenvalues in a Pöschel

Teller potential. The function yields an analytical expression for $|b(k)|^2$. The variables Transmission and Reflection contain the expressions for the transmission and the reflection coefficients. w1a and w2a contain the approximations for $x \to -\infty$ and $x \to \infty$, respectively.

PoeschelTeller PoeschelTeller[x_, n_, N_] calculates the eigenfunction of the Pöschel Teller potential for discrete eigenvalues. N determines the depth of the potential $V_0 Sech(x)$ by $V_0 = N(N+1)$. n fixes the state where $0 < n \le N$.

PlotPT PlotPT[kini_,kend_,type_] gives a graphical representation of the reflection or transmission coefficient depending on the value of the variable type. If type is set to the string r the reflection coefficient is plotted. If type is set to t the transmission coefficient is represented. This function creates 5 different curves.

Reflection Variable containing the reflection coefficient. The independent variables are N and k.

Transmission Variable containing the expression for the transmission coefficient. The independent variables are N and k.

w1a The variable contains the analytical expression for the asymptotic approximation for $x \to -\infty$.

w2a The variable contains the analytical expression for the asymptotic approximation for $x \to \infty$.

boundary.m   Boundary value problem of electrodynamics.

Potential Potential[boundary_,R_,alpha_,n_] calculates the potential in a circular segment. Input parameters are the potential on the circle, the radius R of the circle and the angle of the segment of the circle. The last argument n determines the number of expansion terms used to represent the solution.

centralf.m   Quantum mechanical description of motion in a central field.

Angle Angle[theta_, phi_, l_, m_] calculates the angular part of the wave function for an electron in the Coulomb potential. The numbers L and m denote the quantum numbers for the angular momentum operator. *theta* and *phi* are the angles in the spherical coordinate system.

AnglePlot AnglePlot[pl_,theta_,phi_] gives a graphical representation of the function contained in *pl*. The range of representation is $\pi \le phi < 5\pi/2$ and $0 < theta < \pi$. *Theta* is measured with respect to the vertical axis. This function is useful for plotting the orbitals and the angular part of the eigenfunction.

Orbital Orbital[theta_,phi_,l_,m_,type_String] calculates the superposition of two wave functions for the quantum numbers $m_l = +m$ and $m_l = -m$. The variable type allows the creation of the sum or the difference of the wave functions. The string values of type are either plus or minus.

Radial Radial[ro_, n_, l_, Z_] calculates the radial representation of the eigenfunctions for an electron in the Coulomb potential. The numbers n and l are the quantum numbers for the energy and the angular momentum operator. Z specifies the number of charges in the nucleus. The radial distance between the center and the electron is given by ro.

harmonic.m The harmonic oscillator in quantum mechanics.

a a[psi_, xi_:x] annihilation operator for eigenfunction psi. The second argument specifies the independent variable of the function psi.

across across[psi_, xi_:x] creation operator for eigenfunction psi. The second argument specifies the independent variable of psi.

Psi Psi[xi_,n_] represents the eigenfunction of the harmonic oscillator. The first argument xi is the spatial coordinate. The second argument n fixes the eigenstate.

wcl wcl[xi_,n_] calculates the classical probability of locating the particle in the harmonic potential. The first argument xi is the spatial coordinate while n determines the energy given as the eigenvalue.

wqm wqm[xi_,n_] calculates the quantum mechanical probability for an eigenvalue state n. The first argument represents the spatial coordinate.

kdvanaly.m Multi–soliton solution of the Korteweg-de Vries equation.

Soliton Soliton[x_,t_,N_] creates the N soliton solution of the KdV equation.

PlotKdV PlotKdV[tmin_,tmax_,dt_,N_] calculates a sequence of pictures for the N soliton solution of the KdV equation. The time interval of the representation is $[tmin, tmax]$. The variable dt measures the length of the time step.

kdvequat.m Derivation of the Korteweg-de Vries equation.

Equation Equation[n_] calculates the evolution equation up to order n.

kdvinteg.m Integral of motion of the Korteweg-de Vries equation.

Gardner Gardner[N_] calculates the densities of the integrals of motion for the KdV equation using Gardner's method. The integrals are determined up to the order N.

kdvnumer.m Numerical solution of the Korteweg-de Vries equation.

KdVNIntegrate KdVNIntegrate[initial_,dx_,dt_,M_] carries out a numerical integration of the KdV equation using the procedure of Zabusky & Kruskal. The input parameter initially determines the initial solution in the procedure; e.g., $-6Sech^2[x]$. The infinitesimals dx and dt are the steps with respect to the spatial and temporal directions. M fixes the number of steps along the x–axis.

koch.m Fractal curves.

Fractal Fractal[curve_String, options___] creates a graphical representation of a fractal curve. The type of curve is determined by the first argument. A list of available curves is obtained by calling Fractal[List] or Fractal[Help]. The second argument allows changing the options of the function. The default values are $Generations \rightarrow 3$, $Angle \rightarrow Pi/6$ and $Corners \rightarrow 6$.

lightben.m The bending of a light beam near a planet is discussed.

Deviation Deviation[radius_,mass_] calculates the numerical value of the light bending in a gravitational field of a star with mass M in a distance radius of the center.

Orbit Orbit[radius_,mass_] plots the orbit of a light beam near a mass in the distance radius. The calculation is done in Schwarzschild metric.

linearc.m Motion in a linear chain of particles.

ChainPlot ChainPlot[Initial_List,tend_] creates a plot of the chain for 10 time-steps. Input variables are a list of coordinates and velocities for the initial time $t0$. The second argument is the end point of the time. As output we get 10 pictures stored in the variables $pl[1] - pl[10]$.

membrane.m Vibrations of a circular membrane.

Evolution Evolution[h1_,h2_,g1_,g2_,t1_,order_,number_] calculates the vibration of a membrane for the initial conditions $h1$, $h2$, $g1$ and $g2$ at time $t1$. The input variable *order* determines the order of the Bessel function while *number* determines the number of roots of the Bessel function. The solution is represented by a surface plot in three dimensions.

TimeSeries TimeSeries[t_] plots a sequence of states of the vibrating membrane. This function uses the analytical expression derived by the function Evolution. The function $Evolution[]$ is used prior to using $TimeSeries[]$.

multifra.m Multi-fractal properties of point sets.

MultiFractal MultiFractal[p_List,r_List] calculates the multi-fractal spectrum D_q for a model based on the probabilities p and the scaling factors r. This function plots five functions $Tau(q)$, $D\_q(q)$, $Alpha(q)$, $f(q)$ and $f(Alpha)$.

penning.m Motion of two ions in a Penning trap.

PenningCMPlot PenningCMPlot[x0_,y0_,x0d_,y0d_,w_] gives a graphical representation of the center of mass motion for two ions in the Penning trap. The plot is created for a fixed cyclotron frequency w in Cartesian coordinates (x, y, z). $x0$, $y0$, $x0d$ and $y0d$ are the initial conditions for integration.

PenningI PenningI[r0_,z0_,e0_,n_,l_,te_] determines the numerical solution of the equation of motion for the relative components. To integrate the equations of motion, the initial conditions $r0 = r(t = 0)$, $z0 = z(t = 0)$, and the total energy $e0$ are needed as input parameters. The momentum with respect to the r direction is set to $pr0 = 0$. Parameters l and n determine the shape of the potential. The last argument te specifies the end point of the integration.

periheli.m Perihelion shift of a planet.

AngularMomentum AngularMomentum[minorAxes_,majorAxes_,mass_] calculates the angular momentum of a planet.

D0Orbit D0Orbit[planet_String,phiend_,options___] plots the orbit in the case of vanishing determinants (see text).

Energy Energy[minorAxes_,majorAxes_,mass_] calculates the energy of a planet.

orbit orbit[phiend_,minorAxes_,majorAxes_,mass_] creates a graphical representation of the perihelion shift if the major and minor axes and the mass are given.

Orbit Orbit[planet_String] creates a graphical representation of the perihelion shift for the planets contained in the data base.

PerihelionShift PerihelionShift[minorAxes_,majorAxes_,mass_] Calculates the numerical value of the perihelion shift.

Planets Planets[planet_String] creates a list of data for planets and planetoids stored in the data base of the package PerihelionShift. The data base contains the names of the planets, their major axes, their eccentricities and the mass of the central planet. *Planets*[*'List'*] creates a list of the planets in the data base. *Planets*[*'name'*] delivers the data of the planet given in the argument.

pointcha.m Fields of point charges.

EnergyDensity EnergyDensity[coordinates_List] calculates the density of the energy for an ensemble of point charges. The Cartesian coordinates are lists in the form of $\{\{x, y, z, charge\}, \{...\}, ...\}$.

Field Field[coordinates_List] calculates the electric field for an ensemble of point charges. The Cartesian coordinates are lists in the form $\{\{x, y, z, charge\}, \{...\}, ...\}$.

FieldPlot FieldPlot[coordinates_List,type_,options__] creates a *ContourPlot* for an ensemble of point charges. The plot type (Potential, Field, or Density) is specified as a string in the second input variable. The third argument allows a change of the Options of ContourPlot and PlotGradientField.

Potential Potential[coordinates_List] creates the potential of an assembly of point charges. The Cartesian coordinates of the locations of the charges are given in the form of $\{\{x, y, z, charge\}, \{x, y, z, charge\}, ...\}$.

poisson.m Canonical Poisson bracket.

PoissonBracket PoissonBracket[a_, b_, q_List, p_List] calculates the Poisson bracket for two functions a and b which depend on the variables p and q. Example: *PoissonBracket*$[q, p, \{q\}, \{p\}]$ calculates the fundamental bracket relation between the coordinate and momentum.

quantumw.m Quantum well in one dimension.

PsiASym PsiASym[x_,k_,a_] determines the antisymmetric eigenfunction for a potential well of depth $-V0$. The input parameter k fixes the energy and $2a$ the width of the well. *PsiASym* is useful for a numerical representation of eigenfunctions.

PsiSym PsiSym[x_,k_,a_] determines the symmetric eigenfunction for a potential well of depth $-V0$. The input parameter k fixes the energy and $2a$ the width of the well. *PsiSym* is useful for a representation of eigenfunctions.

Spectrum Spectrum[V0_,a_] calculates the negative eigenvalues in a potential well. $V0$ is the potential depth and $2a$ the width of the well. The eigenvalues are returned as a list and are available in the variables *lsym* and *lasym* as replacement rules. The corresponding eigenfunctions are stored in the variables *Plsym* and *Plasym*. The determining equation for the eigenvalues is plotted.

renormal.m Renormalization and percolation.

Dim Dim[n_] calculates the fractal dimension for the critical probability p_c. The dimension depends on m where $1 \leq m \leq n - 2$; n is the size of the block used.

Nc Nc[n_] determines the mean number of atoms at the probability p_c if m is changed in the range $1 \leq m \leq n - 2$. The size of the block is determined by n.

Pcrit Pcrit[n_] determines the critical probability for an $n \times n$ grid under the variation of m, where m is the number of empty locations in the grid. The range of m is $1 \leq m \leq n - 2$.

RenormPlot RenormPlot[n_,type_String] plots the functions Nc, Dim or $Pcrit$.

tree.m Fractal tree.

Tree Tree[options___] creates a fractal tree. The options of the function Tree determine the form of the fractal created. Options are $Generation \rightarrow 10$, $BranchRotation \rightarrow 0.65$, $BranchSkaling \rightarrow 0.75$, $BranchThickness \rightarrow 0.7$, $OriginalThickness \rightarrow 0.07$, $BranchColor \rightarrow RGBColor[0,0,0]$ Example: $Tree[BranchColor \rightarrow ll, BranchRotation \rightarrow 0.3]$ ll is a list created in the package Tree.

Notebooks

dform.ma Dynamical formulation of equations of motion.

grt.ma General relativity; Einstein's field equations.

intro.ma Introduction to *Mathematica*.

kepler.ma The two body problem.

npend.ma Chaotic behavior of a nonlinear damped driven pendulum.

preface.ma Preface of the book Mathematica in Theoretical Physics.

A.3 *Mathematica* functions

This appendix contains a short description of the *Mathematica* functions used in the book. It is a small selection of the approximately 1200 functions available in the *Mathematica* kernel. The description given does not replace the text of the handbook by S. Wolfram [1.1].

The first few items describe the use of the short hand notation of symbols frequently used in the programming examples. The *Mathematica* function used in the programs and in the notebooks follow.

lhs = rhs evaluates rhs and assigns the result to lhs. From then on, lhs is replaced by rhs whenever it appears. {l1, l2, ... } = {r1, r2, ... } evaluates the ri, and assigns the results to the corresponding li.

lhs -> rhs represents a rule that transforms lhs to rhs.

expr /. rules applies a rule or list of rules to transform each subpart of an expression expr.

lhs := rhs assigns rhs to be the delayed value of lhs. rhs is maintained in an unevaluated form. When lhs appears, it is replaced by rhs, evaluated afresh each time.

lhs :> rhs represents a rule that transforms lhs to rhs, evaluating rhs only when the rule is used.

lhs == rhs returns True if lhs and rhs are identical.

expr //. rules repeatedly performs replacements until expr no longer changes.

AppendTo[s, elem] appends elem to the value of s, and resets s to the result.

Apply[f, expr] or f @@ expr replaces the head of expr by f. Apply[f, expr, levelspec] replaces heads in parts of expr specified by levelspec.

ArcSin[z] gives the arc sine of the complex number z.

ArcTan[z] gives the inverse tangent of z. ArcTan[x, y] gives the inverse tangent of y/x where x and y are real, taking into account the quadrant in which the point (x, y) is located.

Begin["context'"] resets the current context.

BeginPackage["context'"] makes context' and System' the only active contexts. BeginPackage["context'", {"need1'", "need2'", ... }] calls Needs on the needi.

BesselJ[n, z] gives the Bessel function of the first kind J(n, z).

Block[{x, y, ... }, expr] specifies that expr is to be evaluated with local values for the symbols x, y, ... Block[{x = x0, ... }, expr] defines initial local values for x, ... Block[{vars}, body /; cond] allows local variables to be shared between conditions and function bodies.

C[i] is the default form for the i-th constant of integration produced in solving a differential equation with DSolve.

Chop[expr] replaces approximate real numbers in expr that are close to zero by the exact integer 0. Chop[expr, tol] replaces with 0 approximate real numbers in expr that differ from zero by less than tol.

Circle[{x, y}, r] is a two-dimensional graphics primitive that represents a circle of radius r centered at the point {x, y}. Circle[{x, y}, {rx, ry}] yields an ellipse with semi-axes rx and ry. Circle[{x, y}, r, {theta1, theta2}] represents a circular arc.

Clear[symbol1, symbol2, ...] clears values and definitions of the specified symbols. Clear["pattern1", "pattern2", ...] clears values and definitions of all symbols whose names match any of the specified string patterns.

Coefficient[expr, form] gives the coefficient of form in the polynomial expr. Coefficient[expr, form, n] gives the coefficient of form ñ in expr.

ContourPlot[f, {x, xmin, xmax}, {y, ymin, ymax}] generates a contour plot of f as a function of x and y.

Cos[z] gives the cosine of z.

Cosh[z] gives the hyperbolic cosine of z.

Cot[z] gives the cotangent of z.

D[f, x] gives the partial derivative of f with respect to x. D[f, {x, n}] gives the nth partial derivative with respect to x. D[f, x1, x2, ...] gives a mixed derivative.

expr[[i]] or Part[expr, i] gives the ith part of expr. expr[[-i]] counts from the end. expr[[0]] gives the head of expr. expr[[i, j, ...]] or Part[expr, i, j, ...] is equivalent to expr[[i]] [[j]] ... . expr[[{i1, i2, ... }]] gives a list of the parts i1, i2, ... of expr.

f' represents the derivative of a function f of one argument. Derivative[n1, n2, ...][f] is the general form, representing a function obtained from f by differentiating n1 times with respect to the first argument, n2 times with respect to the second argument, and so on.

Det[m] gives the determinant of the square matrix m.

Disk[{x, y}, r] is a two-dimensional graphics primitive that represents a filled disk of radius r centered at the point {x, y}. Disk[{x, y}, {rx, ry}] yields an elliptical disk with semi-axes rx and ry. Disk[{x, y}, r, {theta1, theta2}] represents a segment of a disk.

Display[channel, graphics] writes graphics or sound to the specified output channel.

Do[expr, {imax}] evaluates expr imax times. Do[expr, {i, imax}] evaluates expr with the variable i successively taking on the values 1 through imax (in steps of 1). Do[expr, {i, imin, imax}] starts with i = imin. Do[expr, {i, imin, imax, di}] uses steps di. Do[expr, {i, imin, imax}, {j, jmin, jmax}, ...] evaluates expr looping over different values of j, etc. for each i. Do[] returns Null, or the argument of the first Return it evaluates.

DSolve[eqn, y[x], x] solves a differential equation for the functions y[x], with independent variable x. DSolve[{eqn1, eqn2, ... }, {y1[x1, ...], ... }, {x1, ... }] solves a list of differential equations.

Dt[f, x] gives the total derivative of f with respect to x. Dt[f] gives the total differential of f. Dt[f, {x, n}] gives the nth total derivative with respect to x. Dt[f, x1, x2, ...] gives a mixed total derivative.

EllipticK[m] gives the complete elliptic integral of the first kind K(m).

End[ ] returns the present context, and reverts to the previous one.

EndPackage[ ] restores $Context and $ContextPath to their values before the preceding BeginPackage, and prefixes the current context to the list $ContextPath.

Evaluate[expr] causes expr to be evaluated, even if it appears as the argument of a function whose attributes specify that it should be held unevaluated.

Exp[z] is the exponential function.

Expand[expr] expands products and positive integer powers in expr. Expand[expr, patt] avoids expanding elements of expr which do not contain terms matching the pattern patt.

FindRoot[lhs == rhs, {x, x0}] searches for a numerical solution to the equation lhs == rhs, starting with x == x0.

Flatten[list] flattens out nested lists. Flatten[list, n] flattens to level n. Flatten[list, n, h] flattens subexpressions with head h.

Floor[x] gives the greatest integer less than or equal to x.

FontForm[expr, {"font", size}] specifies that expr should be printed in the specified font and size.

Function[body] or body& is a pure function. The formal parameters are # (or #1), #2, etc. Function[x, body] is a pure function with a single formal parameter x. Function[{x1, x2, ... }, body] is a pure function with a list of formal parameters. Function[{x1, x2, ... }, body, {attributes}] has the given attributes during evaluation.

<<name reads in a file, evaluating each expression in it, and returning the last one. Get["name", key] gets a file that has been encoded with a certain key.

Graphics[primitives, options] represents a two-dimensional graphical image.

GraphicsArray[{g1, g2, ... }] represents a row of graphics objects. Graphics-Array[{{g11, g12, ... }, ... }] represents a two-dimensional array of graphics objects.

HermiteH[n, x] gives the nth Hermite polynomial.

Hold[expr] maintains expr in an unevaluated form.

Hue[h] specifies that graphical objects which follow are to be displayed, if possible, in a color corresponding to hue h. Hue[h, s, b] specifies colors in terms of hue, saturation and brightness.

If[condition, t, f] gives t if condition evaluates to True, and f if it evaluates to False. If[condition, t, f, u] gives u if condition evaluates to neither True nor False.

Im[z] gives the imaginary part of the complex number z.

Infinity is a symbol that represents a positive infinite quantity.

Input[ ] interactively reads in one *Mathematica* expression. Input["prompt"] requests input, using the specified string as a prompt.

Integrate[f,x] gives the indefinite integral of f with respect to x. Integrate[f,{x, xmin,xmax}] gives the definite integral. Integrate[f,{x,xmin,xmax},{y,ymin,-ymax}] gives a multiple integral.

InterpolatingFunction[range, table] represents an approximate function whose values are found by interpolation.

JacobiAmplitude[u, m] gives the amplitude for Jacobi elliptic functions.

JacobiSN[u, m] gives the Jacobi elliptic function sn at u for the parameter m.

Join[list1, list2, ...] concatenates lists together. Join can be used on any set of expressions that have the same head.

LaguerreL[n, x] gives the nth Laguerre polynomial. LaguerreL[n, a, x] gives the nth generalized Laguerre polynomial.

LegendreP[n, x] gives the nth Legendre polynomial. LegendreP[n, m, x] gives the associated Legendre polynomial.

Length[expr] gives the number of elements in expr.

lhs == rhs returns True if lhs and rhs are identical.

Limit[expr, x->x0] finds the limiting value of expr when x approaches x0.

Line[{pt1, pt2, ... }] is a graphics primitive which represents a line joining a sequence of points.

{e1, e2, ... } is a list of elements.

ListPlot[{y1, y2, ... }] plots a list of values. The x coordinates for each point are taken to be 1, 2, ... ListPlot[{{x1, y1}, {x2, y2}, ... }] plots a list of values with specified x and y coordinates.

Log[z] gives the natural logarithm of z (logarithm to base E). Log[b, z] gives the logarithm to base b.

Map[f, expr] or f /@ expr applies f to each element on the first level in expr. Map[f, expr, levelspec] applies f to parts of expr specified by levelspec.

MapAt[f, expr, n] applies f to the element at position n in expr. If n is negative, the position is counted from the end. MapAt[f, expr, {i, j, ... }] applies f to the part of expr at position {i, j, ... }. MapAt[f, expr, {{i1, j1, ... }, {i2, j2, ... }, ... }] applies f to parts of expr at several positions.

MatrixForm[list] prints the elements of list arranged in a regular array.

Max[x1, x2, ...] yields the numerically largest of the xi. Max[{x1, x2, ... }, {y1, ... }, ...] yields the largest element of any of the lists.

Min[x1, x2, ...] yields the numerically smallest of the xi. Min[{x1, x2, ... }, {y1, ... }, ...] yields the smallest element of any of the lists.

Mod[m, n] gives the remainder on division of m by n. The result has the same sign as n.

N[expr] gives the numerical value of expr. N[expr, n] does computations to n-digit precision.

NDSolve[eqns, y, {x, xmin, xmax}] finds a numerical solution to the differential equations eqns for the function y with the independent variable x in the range xmin to xmax. NDSolve[eqns, {y1, y2, ... }, {x, xmin, xmax}] finds numerical solutions for the functions yi. NDSolve[eqns, y, {x, x1, x2, ... }] forces a function evaluation at each of x1, x2, ... The range of numerical integration is from Min[x1, x2, ...] to Max[x1, x2, ...].

**Needs["context`", "file"]**  loads file if the specified context is not already in $Packages. Needs["context`"] loads the file specified by ContextToFilename["context`"] if the specified context is not already in $Packages.

Nest[f, expr, n] gives an expression with f applied n times to expr.

NestList[f, expr, n] lists the results of applying f to expr 0 through n times.

NIntegrate[f, {x, xmin, xmax}] gives a numerical approximation to the integral of f with respect to x over the interval xmin to xmax.

Normal[expr] converts expr to a normal expression, from a variety of special forms.

NSolve[eqns, vars] attempts to solve numerically an equation or set of equations for the variables vars. Any variable in eqns but not vars is regarded as a parameter. NSolve[eqns] treats all variables encountered as vars above. NSolve[eqns, vars, prec] attempts to solve numerically the equations for vars using prec digits precision.

Off[symbol::tag] switches off a message, so that it is no longer printed. Off[s] switches off tracing messages associated with the symbol s. Off[m1, m2, ...] switches off several messages. Off[ ] switches off all tracing messages.

On[symbol::tag] switches on a message, so that it can be printed. On[s] switches on tracing for the symbol s. On[m1, m2, ...] switches on several messages m1, m2, ... On[ ] switches on tracing for all symbols.

ParametricPlot[{fx, fy}, {t, tmin, tmax}] produces a parametric plot with x and y coordinates fx and fy generated as a function of t. ParametricPlot[{{fx, fy}, {gx, gy}, ... }, {t, tmin, tmax}] plots several parametric curves.

ParametricPlot3D[{fx, fy, fz}, {t, tmin, tmax}] produces a three-dimensional space curve parameterized by a variable t which runs from tmin to tmax. ParametricPlot3D[{fx, fy, fz}, {t, tmin, tmax}, {u, umin, umax}] produces a three-dimensional surface parameterized by t and u. ParametricPlot3D
[{fx, fy, fz, s}, ...] shades the plot according to the color specification s. ParametricPlot3D[{{fx, fy, fz}, {gx, gy, gz}, ... }, ...] plots several objects together.

Partition[list, n] partitions list into non-overlapping sublists of length n. Partition[list, n, d] generates sublists with offset d. Partition[list, {n1, n2, ... }, {d1, d2, ... }] partitions successive levels in list into length ni sublists with offsets di.

Pi is pi, with numerical value 3.14159... .

Plot[f, {x, xmin, xmax}] generates a plot of f as a function of x from xmin to xmax. Plot[{f1, f2, ... }, {x, xmin, xmax}] plots several functions fi.

x + y + z represents a sum of terms.

Point[coords] is a graphics primitive that represents a point.

x$\hat{\;}$y gives x to the power y.

PowerExpand[expr] expands nested powers, powers of products, logarithms of powers, and logarithms of products. PowerExpand[expr, {x1, x2, ... }] expands expr with respect to the x1. Use PowerExpand with caution because PowerExpand does not pay attention to branch cuts.

Print[expr1, expr2, ...] prints the expri, followed by a newline (line feed).

Protect[s1, s2, ...] sets the attribute Protected for the symbols si. Protect["form1", "form2", ...] protects all symbols whose names match any of the string patterns formi.

Quit[ ] terminates a Mathematica session.

Random[ ] gives a uniformly distributed pseudorandom Real in the range 0 to 1. Random[type, range] gives a pseudorandom number of the specified type, lying in the specified range. Possible types are: Integer, Real and Complex. The default range is 0 to 1. You can give the range {min, max} explicitly; a range specification of max is equivalent to {0, max}.

Re[z] gives the real part of the complex number z.

ReleaseHold[expr] removes Hold and HoldForm in expr.

Replace[expr, rules] applies a rule or list of rules in an attempt to transform the entire expression expr.

expr /. rules applies a rule or list of rules in an attempt to transform each subpart of an expression expr.

expr //. rules repeatedly performs replacements until expr no longer changes.

RGBColor[red, green, blue] specifies that graphical objects which follow are to be displayed, if possible, in the color given.

lhs -> rhs represents a rule that transforms lhs to rhs.

Save["filename", symb1, symb2, ...] appends the definitions of the symbols symbi to a file.

Series[f, {x, x0, n}] generates a power series expansion for f about the point x = x0 to order $(x - x0)\hat{n}$. Series[f, {x, x0, nx}, {y, y0, ny}] successively finds series expansions with respect to y, then x.

Show[graphics, options] displays two- and three-dimensional graphics, using the options specified. Show[g1, g2, ...] shows several plots combined. Show can also be used to play Sound objects.

Simplify[expr] performs a sequence of transformations on expr, and returns the simplest form it finds.

Sin[z] gives the sine of z.

Sinh[z] gives the hyperbolic sine of z.

Solve[eqns, vars] attempts to solve an equation or set of equations for the variables vars. Any variable in eqns but not vars is regarded as a parameter. Solve[eqns] treats all variables encountered as vars above. Solve[eqns, vars, elims] attempts to solve the equations for vars, eliminating the variables elims.

Sort[list] sorts the elements of list into canonical order. Sort[list, p] sorts using the ordering function p.

SphericalHarmonicY[l, m, theta, phi] gives the spherical harmonic Ylm(theta, phi).

Sqrt[z] gives the square root of z.

Sum[f, {i, imax}] evaluates the sum of f with i running from 1 to imax. Sum[f, {i, imin, imax}] starts with i = imin. Sum[f, {i, imin, imax, di}] uses steps di. Sum[f, {i, imin, imax}, {j, jmin, jmax}, ...] evaluates a multiple sum.

Table[expr, {imax}]   generates a list of imax copies of expr. Table[expr, {i, imax}] generates a list of the values of expr when i runs from 1 to imax. Table[expr, {i, imin, imax}] starts with i = imin. Table[expr, {i, imin, imax, di}] uses steps di. Table[expr, {i, imin, imax}, {j, jmin, jmax}, ...] gives a nested list. The list associated with i is outermost.

Take[list, n]   gives the first n elements of list. Take[list, -n] gives the last n elements of list. Take[list, {m, n}] gives elements m through n of list.

Tan[z]   gives the tangent of z.

Text[expr, coords]   is a graphics primitive that represents text corresponding to the printed form of expr, centered at the point specified by coords.

Thread[f[args]]   "threads" f over any lists that appear in args. Thread[f[args], h] threads f over any objects with head h that appear in args. Thread[f[args], h, n] threads f over objects with head h that appear in the first n args. Thread[f[args], h, -n] threads over the last n args. Thread[f[args], h, {m, n}] threads over arguments m through n.

Unprotect[s1, s2, ...]   removes the attribute Protected for the symbols si. Unprotect["form1", "form2", ...] unprotects all symbols whose names textually match any of the formi.

Which[test1, value1, test2, value2, ...]   evaluates each of the testi in turn, returning the value of the valuei corresponding to the first one that yields True.

References

[1]   **Chapter 1**

[1.1]   S. Wolfram, Mathematica: A System for Doing Mathematics by Computer. Addison-Wesley Publ. Comp. Inc., Redwood City, 1991.

[1.2]   M. Abramowitz & I.A. Stegun, Handbook of Mathematical Functions. Dover Publications, Inc., New York, 1968.

[1.3]   N. Blachman, Mathematica: A Practical Approach. Prentice Hall, Englewood Cliffs, 1992.

[1.4]   Ph. Boyland, A. Chandra, J. Keiper, E. Martin, J. Novak, M. Petkovsek, S. Skiena, I. Vardi, A. Wenzlow, T. Wickham–Jones, D. Withoff, and others, Technical Report: Guide to Standard *Mathematica* Packages, Wolfram Research, Inc. 1993.

[2]   **Chapter 2**

[2.1]   R. Maeder, Programming in Mathematica. Addison-Wesley Publ. Comp. Inc., Redwood City, 1991.

[2.2]   L.D. Landau & E.M. Lifshitz, Mechanics. Addison–Wesley, Reading, Massachusetts, 1960.

[2.3]   J. B. Marion, Classical Dynamics of Particles and Systems. Academic Press, New York, 1970.

[2.4]   R. Courant & D. Hilbert, Methods of Mathematical Physics, Vol. 1+2. Wiley (Interscience), New York, 1953.

[3]   **Chapter 3**

[3.1]   G. Arfken, Mathematical Methods for Physicists. Academic Press, New York, 1966.

[3.2]   P.M. Morse & H. Feshbach, Methods of Theoretical Physics. McGraw-Hill, New York, 1953.

[3.3]   W. Paul, O. Osberghaus & E. Fischer, Ein Ionenkäfig. Forschungsbericht des Wissenschafts- und Verkehrsministeriums Nordrhein–Westfalen, **415**, 1, 1958. Similar work has been done by H.G. Dehmelt, Radiofrequency Spectroscopy of Stored Ions I: Storage, Advances in Atomic and Molecular Physics, 3, 53–72, 1967. D.J. Wineland, W.M. Itano, and R.S. van Dyck Jr., High–Resolution Spectroscopy of Stored Ions, Advances in Atomic and Molecular Physics, Vol. 19, 135–186, 1983.

[3.4]   F.M. Penning, Die Glimmentladung bei niedrigem Druck zwischen koaxialen
          Zylindern in einem axialen Magnetfeld. Physica **3**, 873, 1936. Similar work has
          been done by D. Wineland, P. Ekstrom, and H. Dehmelt, Monoelectron Oscil-
          lator, Physical Review Letters, Vol. 31, 1279–1282, 1973.
[3.5]   G. Baumann, The Paul trap: a completely integrable model? Physics Letters,
          A 162, 464, 1992.

[4]   **Chapter 4**

[4.1]   E. Schrödinger, Quantisierung als Eigenwertproblem. Analen der Physik, **79**,
          361, 1926.
[4.2]   N. Rosen & P.M. Morse, On the Vibrations of Polyatomic Molecules. Physical
          Review, **42**, 210, 1932.
[4.3]   G. Pöschel & E. Teller, Bemerkungen zur Quantenmechanik des anharmonis-
          chen Oszillators. Z. Physik, **83**, 143, 1933.
[4.4]   W. Lotmar, Zur Darstellung des Potentialverlaufs bei zweiatomigen Molekülen.
          Z. Physik, **93**, 518, 1935
[4.5]   S. Flügge, Practical Quantum Mechanics I + II. Springer-Verlag, Berlin, 1971.
[4.6]   C. Cohen-Tannoudji, B. Diu & F. Laloë, Quantum Mechanics I + II. John Wiley
          & Sons, New York, 1977.

[5]   **Chapter 5**

[5.1]   F. Calogero & A. Degasperis, Spectral Transform and Solitons: Tools to solve
          and investigate nonlinear evolution equations. North-Holland Publication Com-
          pany, Amsterdam, 1982.
[5.2]   V.A. Marchenko, On the Reconstruction of the Potential Energy from Phases
          of the Scattered Waves. Doklady Akademii Nauk SSSR, **104**, 695, 1955.
[5.3]   R.M. Miura, C. Gardner & M.D. Kruskal. Korteweg-de Vries equation and
          generalizations. II Existence of Conservation Laws and Constants of Motion.
          Journal of Mathematical Physics, **9**, 1204, 1968.
[5.4]   T.R. Taha & M.J. Ablowitz, Analytical and numerical solutions of certain non-
          linear evolution equations. I. Analytical. Journal of Computational Physics, **55**,
          192, 1984.
[5.5]   N.J. Zabusky & M.D. Kruskal, Interactions of 'solitons' in a collisionless plasma
          and the recurrence of initial states. Physical Review Letters, **15**, 240, 1965.
[5.6]   R.E. Crandall, *Mathematica* for the Sciences. Addison-Wesley, New York, 1991.

[6]   **Chapter 6**

[6.1]   W. Rindler, Essential Relativity. Springer Verlag, New York, 1977.
[6.2]   C.W. Misner, K.S. Thorne & J.A. Wheeler, Gravitation. Freeman, San Fran-
          cisco, 1973.
[6.3]   H. Stephani, General relativity: An introduction to the gravitational field. Cam-
          bridge University Press, 1982.
[6.4]   M. Berry, Principles of Cosmology and Gravitation. Cambridge University
          Press, Cambridge, 1976.

[7]   **Chapter 7**

[7.1]   T.W. Gray & J. Glynn, Exploring Mathematics with Mathematica. Addison-Wesley Publ. Comp. Inc., Redwood City, 1991.

[7.2]   T.F. Nonnenmacher, G. Baumann & G. Losa, Self organization and fractal scaling patterns in biological systems. In: Trends in Biological Cybernetics, World Scientific, Singapore, **1**, 65, 1990.

[7.3]   A. Barth, G. Baumann & T.F. Nonnenmacher, Measuring Rényi-dimensions by a modified box algorithm. Journal of Physics A: Mathematical and General, **25**, 381, 1992.

[7.4]   B. Mandelbrot, The fractal geometry of nature. W.H. Freeman a. Comp., New York, 1983.

[7.5]   A. Aharony, Percolation. In: Directions in condensed matter physics (Eds. G. Grinstein & G. Mazenko). World Scientific, Singapore, 1986.

[7.6]   T. Grossman & A. Aharony, Structure and perimeters of percolation clusters. Journal of Physics A: Mathematical and General, **19**, L745, 1986.

[7.7]   P.G. Gennes, Percolation – a new unifying concept. La Recherche, **7**, 919, 1980.

Index

Since this field is fast-moving, we expect updates and changes to occur that might necessitate sending you the most current pertinent information by paper, electronic media, or both, regarding *Mathematica® in Theoretical Physics*. Therefore, in order to not miss out on receiving your important update information, please fill out this card and return it to us promptly. Thank you.

Name: _____

Title: _____

Company: _____

Address: _____

City: _____ State: _____ Zip: _____

Country: _____ Phone: _____

E-mail: _____

Areas of Interest / Technical Expertise: _____

Comments on this Publication: _____

_____

_____

_____

☐ Please check this box to indicate that we may use your comments in our promotion and advertising for this publication.

Purchased from: _____

Date of Purchase: _____

☐ Please add me to your mailing list to receive updated information on *Mathematica® in Theoretical Physics* and other TELOS publications.

☐ I have a   ☐ IBM compatible   ☐ Macintosh   ☐ UNIX   ☐ other

Designate specific model_____

TELOS®

THE
ELECTRONIC
LIBRARY
OF
SCIENCE

Retu

PLEASE TAPE HERE

FOLD HERE